Praise for *Designing Delivery*

"*Designing Delivery* is an excellent resource on the relationship between design, cybernetics, and IT. Sussna lucidly breaks down dualisms that hold back design and development, and argues for a holistic view of value creation. I would recommend this book especially for those involved in service design who seek a deeper understanding of IT systems, and for IT professionals seeking to better understand how their work extends to the outer branches of service ecosystems."

—Thomas Wendt, Author of
Design for Dasein: Understanding the Design of Experiences

"Jeff Sussna''s book is an excellent guide to the deep context within which digital systems are developed and used. Just as Pirsig asked us to evaluate *Quality* in *Zen and the Art of Motorcycle Maintenance*, so Jeff does for the 21st century. I am sure it will be a classic, too."

—Justin Arbuckle, VP EMEA and
Chief Enterprise Architect at Chef

"*Designing Delivery* is a unique portrayal of the Zeitgeist of DevOps in Information Technology for practitioners. Redefining service design as an on-going dialogue, rather than a race to the finish, it elegantly weaves a cultural narrative, strongly rooted in the systemic ideas of cybernetics and promise theory, but always infused with the deep commitment to humanism that Jeff is well known for."

—Mark Burgess, Author of
In Search of Certainty

"In *Designing Delivery*, Jeff Sussna positions an often misunderstood concept of empathy within the rigorous framework of systems thinking, cybernetics, and promise theory. In doing so, he presents a compelling argument on why 'empathy is the starting point as well as the result of the entire process [of IT].' If you're looking to find out what perspective shifts can help you run a successful IT business in the increasingly complex climate of our post-industrial society, this book is designed to deliver."

—Seung Chan Lim (Slim), Principal Meta-Designer and Researcher at Forks & Bridges, Author of *Realizing Empathy: An Inquiry Into the Meaning of Making*

Designing Delivery

Jeff Sussna

Beijing · Boston · Farnham · Sebastopol · Tokyo

Contents

Foreword

In the late 1990s, I had the honor of working at Forte Software, a company that produced a development and operations platform for what at the time passed for "large scale distributed systems." The brainchild of Paul Butterworth—a man I believe gets less credit than he deserves in the annuls of distributed systems history—Forte allowed developers to develop applications in a scalable software services model, and then easily deploy the resulting user interfaces and services.

Auto-scaling was built into the platform, though you had to architect smartly to use it well. Redundancy and failure recovery were also selectable by checking a box and entering the number of instances you wanted of any given service. To this day, I have yet to see another developer environment that provides the experience Forte offered for distributed systems—modern Platform-as-a-Service offerings included.

This was my first experience with an environment that bridged development and operations functions, and—though I didn't recognize it at the time—it created a different dynamic in the teams that built and deployed applications. I remember frequent conversations between the operations, quality assurance, and development teams; three or four people sitting at a computer discussing the best way to organize, test, scale, and replicate services to maximize availability and performance while minimizing cost—both financially and psychologically.

These days, multidisciplinary approaches to software are a best practice—necessary, even, in enabling the extremes of scale we've reached in digital era. We've even created a term to discuss these concepts: DevOps. The work functions of analysis and design, development, quality assurance, and operations are no longer a linear flow from one to the other, but a set of activities that must all be executed in the face of constant change.

Multidisciplinary software practices, to me, are at the heart of what makes the modern information technology possible: the retraining of technologists to understand and empathize with not only the end user, but also the other technologists that must work on or with the system; the breakdown and rebuilding of organizations to reflect the complex systems nature of the software they are creating; the creation of entire industry ecosystems around services, monitoring, tools, and practices that reflect complexity and constant motion.

In 2011, when I began working at Enstratius (later purchased by Dell), my frequent trips to my old stomping grounds in Minneapolis brought me in touch with a gentleman I had only recently met through Twitter. He was extremely curious about cloud computing, and knew of me through my former blog, The Wisdom of Clouds, on CNET. Over coffee, Jeff Sussna and I sat down and discussed what the adoption of cloud computing, and related changes to application development, were having on our respective areas of interest.

For me, who was primarily focused on application development and operations, the effect of continuous deployment, agility, and "lean" principles signaled the importance of multidisciplinary practices. But Jeff introduced me to another facet of what was changing with respect to software practices.

Jeff's background in quality assurance led him to ask questions about how you can possibly ensure quality in an ever-changing and complex application environment. He was intrigued by complex adaptive systems science, and how it was being applied to application design and operations. However, he introduced me to the importance of empathy to assure that the resulting software thrived—rather than dying a lonely death.

The term *design thinking* meant little to me before I started talking to Jeff, but over the intervening years he helped me understand. I had read Norbert Wiener's *Cybernetics*, but Jeff went much farther than I had in applying those concepts to actual software practices. I'd thought about Mark Burgess's promise theory since I first heard about it in 2008 or so, but Jeff placed promises in the context of design with a focus on extracting quality user experiences from complex software systems.

I'm overjoyed that Jeff wrote *Designing Delivery*. Not only because I'm excited to see his ideas and observations collected in a single (and beautifully written) narrative, but also because I think there are no other books that come at the overall problem of digital services development from the same angle. This most certainly is not just another "DevOps," "lean," or "agile" book. Here, the concepts of cybernetics, service design, and promise theory enable the reader to look beyond technical methodologies to understand the motivations that drive end user behavior, and thus design better, higher quality service experiences.

Those experiences will define the winners and losers of the digital services era. As Jeff will show you, the ability of IT to constantly define and improve those experiences will define its own success in the coming decades.

—James Urquhart, Director of Product, Cloud Management at Dell

Preface

In 1987, I took a job at Apple Computer as a Software Quality Assurance engineer. I was part of a 250-person QA organization. We worked on a different floor, and reported to a different VP, from development. Our jobs were mechanical and somewhat boring. Developers didn't have a lot of respect for us. We didn't have that much more respect for ourselves. I remember a fair number of QA engineers getting caught playing games on their computers all day and being fired as a result. I got out of QA and into development as fast as I could.

Not long after I switched departments, Apple dissolved the QA organization and embedded individual QA engineers into their respective development teams. The testers sat next to the developers, went out to lunch with them, and worked cheek-by-jowl with them. Mutual respect, job satisfaction, and quality all skyrocketed.

Since then I've held a variety of jobs across the Development-QA-Operations spectrum. I've been a consultant, a CTO, and an enterprise architect. I've worked for Software-as-a-Service startups, and managed hosting providers and Fortune 500 companies. I've led Agile, ITIL, and DevOps initiatives. Throughout all this time, my experience at Apple has always stuck with me. It taught me that, as much as anything, quality comes from dissolving boundaries, fostering cross-functional collaboration, and treating testing as an integral part of a process rather than something external to it.

Anything we write is at least partly a reflection of ourselves. This book certainly reflects my personal history. My path through IT has been quite nontraditional and nonlinear. As an undergrad, I studied liberal arts, not computer science. I attended an alternative school that stressed interdisciplinary inquiry. Rather than traditional majors, the college organized subjects into four major schools. Most students designed their majors within one of the four schools. Mine crossed three of them!

During my senior year, my advisor encouraged me to take a class that was completely unrelated to anything else I was doing. I took a Lisp programming class, and quickly fell in love with the language. Twenty-five years later, my programming resume includes compilers, websites, and distributed operating systems.

That experience taught me another important lesson. We don't discover solutions to problems purely through analysis. Design is also about adaptation to serendipity. As we move through life and think about the problems we face, we encounter unexpected opportunities. The quality of our solutions depends on our ability to respond to them.

In my case, over the past several years I've stumbled upon three key ideas: design thinking, cybernetics, and promise theory. Together with the lessons I learned from college and from working at Apple, these ideas underpin my approach to software and service quality, and form the conceptual backbone of this book.

The first idea is *design thinking*. As a liberal arts major, I always felt a bit like a stranger in a strange land working in IT. Then I read Tim Brown's book, *Change By Design*. It introduced basic design-thinking principles of empathy, ethnography, divergent thinking, and abduction. Suddenly I had a name for the way I'd always approached technology.

Design thinking introduced me to service design. Service design applies product-design techniques and sensibilities to the design of services. In the era of cloud computing, everything we do with software manifests in the form of services. Service design teaches us that we need to think about Software-as-a-Service in terms of our entire interaction with our customers.

Second, serendipity placed a biography of Norbert Wiener in my path. Wiener coined the term *cybernetics* to refer to feedback-based, adaptive control systems, and was a pioneering member of the field. I'd vaguely heard of the *science of control* before, but knew little about it. Wiener's biography immediately captured my imagination. Since then, cybernetics has come to define my approach to design, development, quality assurance, and operations alike.

Finally, my work with IT operations automation led me to Mark Burgess. Mark is the godfather of system administration theory, and the creator of an infrastructure automation tool called CFEngine. He also conceived a model for describing complex socio-technical systems called promise theory. Promise theory gave me a language to put it all together.

Complex systems such as cloud-based software services are inherently uncertain. Failure is inevitable and unavoidable. Customers, though, expect certainty: in other words, continuous rather than discontinuous quality. How do we reconcile expectations with reality? If we can't avoid failure, can we instead attain resilience?

Promise theory offers a simple yet powerful model for thinking about the relationship between certainty and uncertainty. This model is especially powerful because it lets us address those questions across domains and levels. We can use it to model and manage resilient services as the complex wholes that they are.

Promise theory operates from a cybernetic perspective. Its language gives us a lever to dissolve dualities between acting and listening, designing and operating, and operating and repairing. We can use it to guide our pursuit of what is in my opinion the ultimate goal of twenty-first-century business: dissolving the duality between service provider and consumer.

Systems thinking and transdisciplinary problem solving have always been central to the way I think. I believe they are more than just useful to me personally. They're also crucial to quality in the 21st-century service economy. As digital services become unavoidable components of our daily lives, their quality more and more deeply influences our own quality of life. Unlike products, which engage customers through momentary transactions, service defines ongoing relationships. To create human value through service experiences, we need to create holistic, responsive, and adaptive relationships between companies and customers.

My experience at Apple introduced me to the idea that integrating tools, practices, and perspectives across disciplines leads to improved quality. This book starts from that belief. It follows a path that reflects my naturally integrative way of thinking. The result is an attempt at a profoundly holistic reimagining of the role and purpose of twenty-first-century IT.

In my view, IT's purpose must transform from delivering stable systems and reliable information to enabling responsiveness. Promise theory teaches us that certainty comes from continually repairing uncertainty. Design thinking and cybernetics teach us that the way to repair uncertainty is through user-centered feedback loops. IT quality becomes a measure of how well systems and organizations deliver the capability to respond to customers and markets through continual design.

This definition of quality sees design and operations as inseparable. That view reflects my personal journey, which has crossed liberal arts and engineering. It has led me to a perspective on IT and business that spans technical and nontechnical boundaries. My goal for this book is to create a shared understanding and language that marketers, designers, developers, testers, and operators can use to deepen their collaboration with one another. It is my hope that by helping organizations improve cross-functional collaboration, I can make a small contribution to improving the quality of the services upon which we increasingly depend to conduct business and live our lives.

Safari® Books Online

 *Safari Books Online* is an on-demand digital library that delivers expert *content* in both book and video form from the world's leading authors in technology and business.

Technology professionals, software developers, web designers, and business and creative professionals use Safari Books Online as their primary resource for research, problem solving, learning, and certification training.

Safari Books Online offers a range of plans and pricing for enterprise, government, education, and individuals.

Members have access to thousands of books, training videos, and prepublication manuscripts in one fully searchable database from publishers like O'Reilly Media, Prentice Hall Professional, Addison-Wesley Professional, Microsoft Press, Sams, Que, Peachpit Press, Focal Press, Cisco Press, John Wiley & Sons, Syngress, Morgan Kaufmann, IBM Redbooks, Packt, Adobe Press, FT Press, Apress, Manning, New Riders, McGraw-Hill, Jones & Bartlett, Course Technology, and hundreds more. For more information about Safari Books Online, please visit us online.

How to Contact Us

Please address comments and questions concerning this book to the publisher:

O'Reilly Media, Inc.
1005 Gravenstein Highway North
Sebastopol, CA 95472
800-998-9938 (in the United States or Canada)
707-829-0515 (international or local)
707-829-0104 (fax)

We have a web page for this book, where we list errata, examples, and any additional information. You can access this page at *http://bit.ly/designing-delivery*.

To comment or ask technical questions about this book, send email to *book-questions@oreilly.com*.

For more information about our books, courses, conferences, and news, see our website at *http://www.oreilly.com*.

Find us on Facebook: *http://facebook.com/oreilly*

Follow us on Twitter: *http://twitter.com/oreillymedia*

Watch us on YouTube: *http://www.youtube.com/oreillymedia*

Acknowledgments

The ideas that form the basis for this book started as a series of talks I gave at various technical conferences in 2013 and 2014. During one of those conferences, a conversation with Courtney Nash from O'Reilly Media sparked the notion that it might be possible to weave my thoughts into a coherent narrative. That narrative would never have happened without her intuition that there might be something of interest from which to draw.

Now that I've made it through my first book, I understand why thanking one's family appears prominently in the acknowledgments of every book ever written. Writing is lonely, challenging, and obsessive. At times I wandered the halls of the house like a ghost. Other times I disappeared to coffee shops for entire weekends. I can't thank my wife, Brenda, and my son, Ian enough for their patience, support, and encouragement throughout the process.

I am a compulsive synthesist. This book reflects that fact by bringing together DevOps and associated IT practices with design thinking, a completely "non-technical" methodology. It wasn't until I discovered promise theory, though, that I found a language with which I could fully express this synthesis. I owe a deep debt of gratitude to Mark Burgess for conceiving of promise theory and giving it freely to the world. I also owe him thanks for his kindness, decency, and patience while I grappled with the profundity of his ideas.

Mark generously gave his time to review my manuscript, as did Wim Rampen and James Urquhart. It was James who first introduced me to complexity theory. He was also the first person to be kind to me on Twitter, thus encouraging me to venture further into that world, where I've met so many other helpful, encouraging people.

Chief among them during this project have been Dan Turkenkopf, Graham Hill, Justin Arbuckle, Thomas Wendt, Paul Pangaro, Seung Chan Lim, and Gene Kim, who encouraged me to "fight for the heart of QA". Fredrik Matheson has been an inexhaustible fount of enthusiasm. I also want to thank Patrick Debois and the international DevOps community: first, for giving me repeated opportunities to work through my personal process of synthesis; and second, for being willing to listen to me prattle on about cybernetics and design thinking and promise theory and whatever else came into my mind.

I've always found listening to music helpful during the writing process. Picking the right music for the struggle that accompanies any particular paragraph is a minor art form. Throughout this book, whenever I found myself stuck in a particularly deep rut, the improvisational cello music of Zoe Keating never failed to dissolve my writer's block and pull me out again.

Finally, working with O'Reilly Media has made me a strong believer in the value of a good publisher. Brian Anderson, my editor, deserves credit for helping me turn raw ideas into readable paragraphs and chapters. The rest of the O'Reilly staff, who helped with production, marketing, and speaking, also were tremendously helpful.

Introduction

In 1912, Teddy Roosevelt included a proposal for national health care as part of his presidential campaign platform. Nearly 100 years later, in 2010, Congress passed the Affordable Care Act. The legislative process was highly acrimonious. After the law's passage, the Republican majority in the House of Representatives spent the next three years trying to repeal it. Even now, the law continues to face ongoing legal challenges. The Supreme Court has ruled on its constitutionality once, and likely will be called on to do so again.

The law included two key provisions: citizens could purchase health insurance through state or federal marketplaces, and they would need to buy insurance by the end of 2013 in order to avoid financial penalties. Late in 2013, the government rolled out a website through which people could access the federal marketplace. Its initial release was an unmitigated disaster. The site suffered from pervasive performance and usability problems, security issues, and outages.

Critics immediately blamed a lack of testing. Further exploration, however, uncovered more fundamental issues. Some of those issues were inherent in the nature of government IT procurement. Siloed project management made it difficult for anyone to understand the overall goals or requirements. Various parts of the project were contracted to different vendors. Each vendor worried about its own deliverables. No one took responsibility for integrating the pieces into a working whole. To make matters worse, the Democrats had spread chunks of the project across unrelated budgets in order to hide them from Republican defunding efforts.

No one realized just how critical a website would be to one of the most ambitious and contentious government programs of the past 100 years. The IT vendors involved in the project assumed it would just be another anonymous, behind-the-scenes effort. Government IT projects failed all the time; generally speaking, no one knew or cared about these failures. As a result, no one made the connection between the project's structural problems and the potential severity of their repercussions.

The spectacular fashion in which this failure became public reflects Health-Care.gov's position as the most high-profile example of a defining trend in twenty-first-century society. We are entering the age of the digital service economy. Fewer and fewer of our daily activities happen without some kind of digital component. We interact with these components through services that companies and governments operate on our behalf. Software-as-a-Service is becoming an indispensable part of ordinary life.

This trend shines a very bright spotlight on IT's purpose and its role within the organizations it serves. No longer can we treat IT as an invisible, isolated activity, hidden in the back corner. The quality of the digital systems we build, and the way in which we build and operate them, ascome critical to organizations' ability to function and to serve their customers. When every business becomes a digital business, IT becomes inseparable from the rest of the organization. It becomes the very essence of how businesses operate, and how they interact with customers.

In the post-industrial service economy, customers engage with brands through digital conversations. Brand quality becomes inseparable from operational quality. People's experiences with HealthCare.gov, for example, have tainted their opinions about the Affordable Care Act as a whole. IT thus needs to transform itself from a tool for operational efficiency to a medium that enables high-quality conversations between companies and customers.

The HealthCare.gov fiasco illustrates the need to broaden and deepen our definition of IT quality. That definition, and the techniques we use to achieve it, must reach further than just software functionality. They need to become an integral part of design, development, and operations. They need to concern themselves with service, not just software. Ultimately, our understanding of digital service quality must transcend the confines of IT, and become a mirror that helps the entire service organization empathize with its customers in order to help them solve the problems they face in their daily lives.

Customers judge service quality based on the entirety of their experience. They view usability, functionality, and operability as inseparable dimensions of service. They expect to be able to engage with services coherently over time, across multiple touchpoints. They expect digital brands to deliver services that:

- Address the customer's entire journey across all its touchpoints

- Are available whenever and wherever users need them

- Adapt over time to meet users' changing needs

- Improve over time by learning from failure

HealthCare.gov's initial release didn't fail just because it didn't function well. It failed because it broke its promise to its users. That promise wasn't just to let people find and purchase health insurance. It was to help them navigate a situation of urgency, stress, and confusion. Given the contentious environment in which the website was launched, combined with the deep impact of health insurance on people's lives, the site's implicit brief included the need to educate, calm, and reassure. Its design, functional, and operational problems all contributed to its failure to keep that promise.

Users don't care about websites or IT systems. They care about their own goals. They hire services to co-create value by helping them accomplish those goals. People don't log on to HealthCare.gov or complain when it doesn't work because they want to sign up for health insurance. They do it because they want to be able to run a small business without worrying about being able to afford medical care if they get sick. Or perhaps they just want to minimize their costs and avoid federal penalties. In either case, their goals involve important, emotionally loaded personal needs such as health and finance.

The true definition of quality isn't how well the software works but rather how well the service helps its customers accomplish their practical goals and satisfy their emotional needs. Service providers must address quality across all dimensions of service delivery, including the internal processes by which that delivery happens. The entire organization must align itself with users' goals. Only when it has a holistic understanding of itself and its relationship with its customers can an organization successfully co-create value with them. The government's fractured approach to HealthCare.gov illustrates what happens in the absence of shared understanding and cross-functional collaboration.

Twenty-first-century service providers face a conundrum in the quest to help customers satisfy essential personal needs. On the one hand, they make promises to their customers about providing continuous quality. On the other hand, the human and technical systems they use to deliver these promises are increasingly complex, uncertain, and failure-prone. Digital service organizations must understand how to deliver certainty on top of uncertainty. Quality becomes a matter of resilience rather than stability. Ensuring quality becomes a process of managing the relationship between certainty and uncertainty. It requires the ability to continually achieve success in the face of failure.

The need to continually transform failure into success is not restricted to technical systems. The design of services is a process of continual repair. No matter how well we design something, we can't fully judge its usefulness or even how it will be used until real customers engage with it. Co-creation implies that users contribute to defining solutions through using them. In the process, they generate new problems for designers to solve.

A service provider's ultimate promise is not just to deliver a specific set of capabilities but rather to continually design them for and with customers. When brands become digital conversations, design and operations merge into a unified, never-ending cycle. Especially given the increasingly disruptive nature of the global economy, the ability to continually adapt to changing customer expectations becomes ever more critical to brand quality. Continual co-design with customers also implies continual organizational self-design. IT's new mandate is thus to maximize the fluidity of both internal and external conversations.

This book introduces a new, transdisciplinary approach to digital service that can help companies improve customer satisfaction and create positive brand experiences. It shows readers how to unify their approach to quality across front-office and back-office functions, and technical and business perspectives. It helps IT leaders understand how to transform their role within the organizations they serve. Conversely, it helps other business leaders understand how to engage with IT, and what to expect from it.

The book presents the need for an integrated, continuous approach to brand and IT quality, and the path to achieving it, in three parts:

- Part I describes the transformative effect of twenty-first-century, post-industrial society on business, and in turn on IT.

- Part II presents a new definition of quality that reflects the needs of post-industrial businesses.

- Part III introduces a unified approach to designing and operating services that can help twenty-first-century businesses maximize the quality of their digital conversations.

At its heart, service quality is about making and keeping promises to customers. Service promises cross disciplines and organizational silos. Keeping them requires shared understanding and collaboration between marketing, design, development, QA, operations, support, and management. Service providers need a unified way to understand the explicit and implicit promises they make. They need a way to measure themselves against those promises and to maximize their ability to keep them.

Continuous design is a methodology for identifying, keeping, and repairing service promises. It unifies practices and perspectives from cybernetics, service design, and promise theory. By bringing together these techniques, continuous design offers a coherent approach to achieving continuous quality across all its dimensions.

Cybernetics is the science of adaptive, feedback-based control. The name comes from the Greek word for *steersman*. Cybernetics takes the view that control in complex environments must be conversational. It requires not just action but also listening and adaptation. To steer a boat across a lake, you can't just ignore changing winds and currents; instead, you have to use your tiller and sails to adjust to them. The cybernetic model of control is thus circular: decisions about quality, success, and where to go next depend not just on how well people carry out their intentions but also on how the environment responds to them.

The cybernetic model of control underlies modern IT methodologies such as Agile, DevOps, cloud, and LeanUX. When combined within one another, these methodologies transform IT into a conversational medium that allows businesses to steer in response to continually evolving customer needs and market demands. Through self-steering, they can continually change and adapt while at the same time maintaining their brand's essential identity.

Service design applies product design techniques to the design of everything from services to organizational processes. Its fundamentally user-centered approach redefines quality in terms of desired outcomes rather than features. Its emphasis on holistic interactions over time and across touchpoints emphasizes the view of service delivery as a co-creative journey. Its grounding in design as a creative practice enables exploration and discovery.

Service design expands the scope of design beyond being purely concerned with user interaction. In the process, it helps create customer-focused alignment throughout service organizations. It also raises design's horizon to address the service delivery process itself as a first-class design problem. By doing so, it lets IT organizations continually improve quality by adapting their technologies and practices in response to organizational and market dynamics.

Promise theory describes human and technical systems as collections of autonomous agents that collaborate by making promises to one another. The use of the word *promise* represents the uncertain nature of complex systems. By their nature, promises aren't always kept; sometimes they get broken. Making uncertainty explicit allows us to account for it, and ironically, to achieve greater certainty as a result.

The word *promise* also emphasizes the centrality of service and relationships. We make promises about things we believe are important to beneficiaries about whom we care. We collaborate as necessary in order to keep our promises. Promise theory thus encourages a focus on beneficial outcomes rather than organizational structures or implementation details.

Promise theory directly relates to brand quality. A brand represents a promise to help a customer achieve a desired outcome. People's opinions of a brand reflect the history of the promises it has made, kept, and broken. Promise theory provides a concrete mechanism for understanding and managing the promises that actualize a company's brand. Its unified view of certainty and uncertainty helps brands approach promise keeping as a process of continual learning and repair.

Cybernetics, service design, and promise theory intersect at the point of empathy. In order to accurately steer your response in conversations with customers, it's necessary to understand the motivations driving their side of the conversation. In order to design truly user-centered systems and services, you must respect the integrity of the user's perspective as different from your own. In order to make the right promises at the right times, you must understand which outcomes are meaningful and why customers care about them.

Empathy as the ultimate driver of brand quality is the fundamental theme of this book. Empathy allows service organizations to see themselves from their customers' perspective. It shifts their emphasis from attachment to their own products to attachment to how useful they are in their customers' eyes. It shifts their conception of their underlying business process from one of creation, delivery, and management to one of listening, responding, and adaptation.

Continuous, empathic design is the most fundamental competency that twenty-first-century businesses need. Quality in the service economy is a dynamic, relational process. It depends on the ability to continuously deliver customer benefit. Complexity and disruption generate ever-shifting markets and user needs. Service delivery thus becomes inseparable from service and organizational design.

In order to converse with customers and the market, digital businesses need to apply continuous design to themselves as well as the services they provide. Co-creative service value depends on internal as well as external adaptability. In order to keep customer promises, the components of a digital service organization also must keep promises to one another. Design becomes operations, and operations becomes design.

IT's new mandate thus includes empowering organizations to continuously redesign themselves. The cybernetic, circular nature of digital service implies that customer and employee relationships continually reflect one another. The methodology presented by this book strives to help organizations pursue inner and outer quality as mutually reinforcing aspects of a seamless whole.

This circular interdependency also applies to IT itself. In order to transform itself into a medium for useful customer conversations, IT needs to rethink its role as a delivery mechanism. Enabling the delivery of continuous design becomes its core purpose. In order to fulfill that purpose, IT also needs to continuously design itself. Like any other part of a post-industrial organization, IT needs to develop the ability to adapt not just what it does but also how it does it.

PART I

Post-Industrial IT

From Industrialism to Post-Industrialism

In 1973, Daniel Bell published a book called *The Coming of Post-Industrial Society*. In it, he posited a seismic shift away from industrialism toward a new socioeconomic structure which he named *post-industrialism*. Bell identified four key transformations that he believed would characterize the emergence of post-industrial society:

- Service would replace products as the primary driver of economic activity
- Work would rely on knowledge and creativity rather than bureaucracy or manual labor
- Corporations, which had previously strived for stability and continuity, would discover change and innovation as their underlying purpose
- These three transformations would all depend on the pervasive infusion of computerization into business and daily life

If Bell's description of the transition from industrialism to post-industrialism sounds eerily familiar, it should. We are just now living through its fruition. Every day we hear proclamations touting the arrival of the service economy. Service sector employment has outstripped product sector employment throughout the developed world, according to World Bank (*http://bit.ly/ 1K6vXiG*).

Companies are recognizing the importance of the customer experience. Drinking coffee has become as much about the café and the barista as about the coffee itself. Owning a car has become as much about having it serviced as about driving it. New disciplines such as service design are emerging, which use design techniques to improve customer satisfaction throughout the service experience.

Disruption is driving 100-year-old, blue-chip companies out of business. Startups that rethink basic services like hotels and taxis are disrupting entire industries. *Innovate or die* is the business mantra of the day. Companies are realizing the need to empower their employees' creative decision-making abilities, and are transforming their organizational structures and communications tools in order to unleash internal innovation and collaboration.

IT is moving beyond playing a supporting role in business operations; it's becoming inseparable from it, to the point where every business is becoming a digital business. One would expect a company that sells heating and ventilation systems to specialize in sheet aluminum and fluid dynamics. You couldn't imagine a more *physical-world* industry. Yet HVAC suppliers have begun enabling their thermostats with web access in order to generate data for analytics engines that automatically fine-tune heating and cooling cycles for their customers. As a result, they're having to augment their mechanical engineering expertise with skills in building and running large-scale distributed software systems.

From Products to Service

The Industrial Age focused on optimizing the production and selling of products. Interchangeable parts, assembly lines, and the division of labor enabled economies of scale. It became possible to manufacture millions of copies of the same object. Modern marketing evolved to convince people to buy the same things as one another. Consumerism brought into being a world where people evaluated their lives by what they had, rather than how they felt.

A product economy functions in terms of transactions. A sneaker company, for example, calculates how many units they think they can sell, at what price point, to whom. They create a marketing campaign to drive the desired demand. They link their production and distribution systems to that forecast.

The consumer, for her part, comes home from a run with sore feet and decides she needs a new pair of running shoes. She goes to the local athletic store and tries on a few different kinds of shoes. At some point, she makes a decision and buys something, at which time the transaction is concluded.

In addition to being transaction-oriented, a product economy relies on a push marketing model. Companies use the Four P's (Production, Price, Promotion, and Place) to treat marketing like an "industrial production line that would automatically produce sales."[1] According to this model, proper planning almost predestines the customer to drive to a certain store, try on certain shoes, and make a certain purchase.

The twentieth-century media model developed hand in hand with consumerist marketing. Broadcast television evolved as the perfect medium for companies to try to convince consumers that they needed a particular product. The very words *broadcast* and *consumer* reveal the nature of the relationship between industrial production and marketing.

A post-industrial economy shifts the focus from selling products to helping customers accomplish their goals through service. Whereas products take the form of tangible things that can be touched and owned, service happens through intangible experiences that unfold over time across multiple touchpoints. Consider the example of flying from one city to another. The experience begins when you purchase your ticket, either over the phone or via a website. It continues when you arrive at the airport, check your bags, get your boarding pass, find your gate, and wait for boarding to begin.

Only after you board does the flight actually begin. You buy a drink and watch a movie. Finally the plane lands; you still have to disembark and collect your luggage. The actual act of flying has consumed only a small part of the overall trip. In the process of that trip, you have interacted with ticket agents, baggage handlers, boarding agents, and flight attendants. You have interfaced with telephones, websites, airport signage, seating areas, and video terminals.

Unlike product sales, which generate transactions, service creates continuous relationships between providers and customers. People don't complain about an airline on Twitter because their flight was delayed. They complain because *as usual* their flight was delayed. Perhaps instead they remark on the fact that, for once, their flight wasn't delayed.

Service transforms the meaning of *value*. A product-centric perspective treats value as something to be poured into a product, then given to a customer in exchange for money. If I buy a pair of sneakers but leave them in my closet and never wear them, I don't feel entitled to ask for my money back.

Service value only fully manifests when the customer uses the service. The customer *co-creates* value in concert with the service provider. The fact that an airline owns a fleet of airplanes and sells you a ticket for a seat on one of them doesn't by itself do you any good. The value of the service can't be fully realized

---

1 Wim Rampen, email communication to author, July 16, 2014.

until you complete your flight. You and the airline—and its ticket agents, pilots, flight attendants, and baggage personnel—all have to work together in order for the flight to be successful. The goals, mood, situation, and surrounding experiences you bring with you all contribute to the success of the service experience.

Service changes the dynamic between vendor and consumer, and between marketer/salesperson and buyer. In order to help customers accomplish a goal, you need to understand their goals, and what they bring to the experience. In order to do that, you need to be able to listen, understand, and empathize. Service changes marketing from *push* to *pull*.

Marketers are beginning to adapt to this new model. They are recognizing that strategies like content marketing fail to provide sufficient visibility into the mind of the consumer. Some organizations are supplementing content with marketing applications. These applications flip the Four P's on their head, and give marketers meaningful customer insight through direct interaction.

Astute readers will notice the need for an even larger network of collaboration. The airline operates within an airport, which operates within a city. A successful trip therefore also needs help from security agents and road maintenance crews. High-quality services address the larger contexts in which they co-create value with customers.

Sun Country, for example, is a regional airline headquartered in Minneapolis. Those of us who live in Minneapolis know that Minnesota has two seasons: winter and road construction. Sun Country recognizes that road construction can cause driving delays, so warnings are posted on the home page of its website about construction-related delays on routes leading to the airport.

Sun Country understands that, although the roads around the airport are beyond its control, it can still impact the perceived quality of the service experience. If I arrive at the gate late and feeling harried, I'll have less patience for any mistakes on the part of the ticket agent. I'll more likely find fault with the airline, regardless of who's truly at fault.

The Internet and social media are accelerating the transformation of the marketing and sales model by upending the customer-vendor power structure. Customers now have easy access to as much if not more information about service offerings and customer needs than the vendors themselves. Facebook and Twitter instantly amplify positive and negative service experiences. Customer support is being forced out onto public forums. Companies no longer control customer satisfaction discussion about their products. Instead, they are becoming merely one voice among many.

From Discrete to Infused Experiences

It used to be relatively straightforward to know where in your life you were and what you were doing at any point in time. You were either at home or at work, or else driving between them. If you wanted to hang out with friends, you went to the mall. If, on the other hand, you wanted to be alone, you went in your room and shut the door. If you wanted to surf the Internet, you sat down at a desk in front of a computer. If not, you went out for a walk or a drive.

Now, though, the parts of our lives are melding together and infusing one another. Do you go to the coffee shop to chat or to work? Do you use your phone to call your mom, or to upload photos of your cat to Instagram, or to check your office email? Do you go to the library to peruse hardcover books on shelves, or to read ebooks on websites? Do you use your car for transportation, or as an incredibly complicated and expensive online music player? The answer to all of these questions is "yes."

Even in the digital domain, our daily activities and the tools we use to accomplish them are blending together. We use the same phones, laptops, and cloud services to manage personal and business data. We check our Facebook accounts from work, and read our work email at the kitchen table. We use Twitter to maintain both friendships and professional networks.

Digital infusion has fully blossomed. The word *infusion* refers to the fact that computer systems are no longer separate from anything else we do. The digital realm is infusing the physical realm, like tea in hot water. Or, as Paolo Antonelli (*http://bit.ly/1K6wmSr*), Senior Curator of Architecture and Design for the Museum of Modern Art in New York, put it, "We live today, as you know very well, not in the digital, not in the physical, but in the kind of minestrone that our mind makes of the two."

We encounter fewer and fewer situations that are purely physical. When I go shopping for a new refrigerator, I'm likely to read a review of it on my smartphone while I'm looking at it on the showroom floor. IKEA has integrated its paper catalog with its mobile app. If I use my phone to take a picture of an item in the catalog, the app will bring up more detailed, interactive information about the item in question.

Digital infusion means that brick-and-mortar retailers like Sears and IKEA, which traditionally specialized in the in-store experience, now must also offer equally compelling online experiences. To make things even more challenging, customers expect seamless experiences across physical and virtual channels: stores, kiosks, web browsers, tablets, phones, cars, and so on. As a result, companies are having to expand their marketing, design, and technology expertise to bridge the physical and digital domains.

The digital realm has moved beyond an isolated, contained part of our lives to become the underlying substrate upon which we carry out all of our activities and interactions. With the emergence of the Internet of Things, the digital realm is beginning to completely surround us. Our walls have connected thermostats. Our arms have connected watches. Our lawns have connected sprinkler systems. Our cars have connected dashboards.

In order to serve this newly infused world, companies need to undertake equally deep internal transformations. IT used to be a purely internal corporate function. IT might impact internal operations efficiency, but from the consumer's perspective, it remained invisible—literally behind the scenes—like an automobile assembly line. The ease or pain with which employees shared marketing documents, filed expense reports, or tracked vendor purchase orders was of no concern to the customer.

Infusion breaks down the boundaries between internally facing systems of record and externally facing systems of engagement. The relational, continuous, collaborative nature of service means that internal company operations are inseparable from customer service. In a digitally infused business, therefore, IT becomes an integral part of the customer-facing service. In order for a customer to be able to upgrade a service subscription, for example, a public website may need to interact in real time with a back-office enterprise resource planning (ERP) system. If that ERP system is slow or incapable of providing important data to the website, its failures will become visible to the customer.

The virtualization of experience dramatically raises the stakes for digital service quality. Quality becomes that much more important because people depend on digital services for their very ability to function. If I can't transfer money over the Web from my savings account to my checking account, I might bounce my rent check. If the software in my thermostat has a bug, I can't warm up my house on a cold day. If my corporate ERP system goes down, my customers might not be able to log into their accounts.

From Complicated to Complex Systems

Digital infusion changes not just the way we experience things but also the way we organize, construct, and operate them. If the music player in your car isn't working, is it Honda's fault or Pandora's? If you can't watch a video on Friday night, is it Netflix's, or Comcast's, or Tivo's fault? If you're a freelancer, do you work for yourself, your client, or the broker who got you the gig? Is your invoicing data managed by your accounting SaaS provider, the PaaS on top of which they run, or the IaaS on top of which that PaaS runs? If you have a problem with something you bought, do you call the company's customer support line, or do you just post your question or complaint on Facebook or Twitter?

Infusion breaks down familiar boundaries and structures. No longer can customers assume that Honda will transparently manage all of its vendors in order to deliver a working car, or be able to fix problems with any of its parts. Conversely, Honda can no longer assume it controls the communications channels with its customers.

This dissolving of boundaries impacts IT structures as well. If a customer can't log in to your website, is the problem caused by the web server or the mainframe finance system that holds the customer record? The new requirement for interconnectivity between systems of record and systems of engagement complicates network and security architectures. So-called *rogue* or *shadow* IT, where business units procure cloud-based IT services without the participation of a centralized IT department, makes it harder to control or even know which data lives inside the corporate data center and which lives in a cloud provider's data center.

Infusion forces homogeneous, hierarchical, contained systems to become heterogeneous, networked, fluid, and open-ended. In other words, complicated systems become complex ones. People often use the words *complicated* and *complex* interchangeably. When applied to systems, however, they mean very different things, with very different implications for defining and achieving quality. We therefore need to understand the distinction between them.

COMPLICATED SYSTEMS

A complicated system can have many moving parts. A car contains something on the order of 30,000 individual parts. All 30,000 parts, however, don't directly interact with one another. The fuel system interacts with the engine, which interacts with the drivetrain. The fuel system consists of a fuel tank, fuel pump, and carburetor. The carburetor is made up of jets, float bowls, gaskets, and so on.

Complicated systems arrange their components into navigable, hierarchical structures that facilitate understanding and control. Very few of us can fix our own cars anymore. We can still, though, reasonably understand their overall structure. If our car has a flat tire, and the service technician tells us we need a new carburetor, we know enough to suspect that something fishy is going on.

The interactions within complicated systems don't dynamically change. The carburetor doesn't suddenly start directly interacting with the tires. Furthermore, complicated systems behave coherently as wholes. If you're driving your car, and you turn the steering wheel to the left, the entire car goes to the left. One of the doors doesn't decide to wander off in the opposite direction.

COMPLEX SYSTEMS

Complex systems, on the other hand, consist of large numbers of relatively simple components that have fluid relationships with many other components. Examples of complex systems include everything from ant colonies to companies to cities to economies. You can't really define an economy, for example, as a neat hierarchy, with the Chair of the Federal Reserve at the top, the Fortune 500 below that, and small businesses and sole proprietors at the bottom.

Instead, companies and individuals dynamically create and dissolve business relationships with one another on multiple levels. I hire a plumber. A large company buys a smaller one. An executive quits their position to found a startup competitor. Toyota buys parts from many different vendors. A battery manufacturer, on the other hand, may supply batteries to Toyota, Ford, and BMW.

Complex systems function more like an ongoing dance, with the dancers changing partners on the fly. Schools of fish and flocks of birds offer compelling illustrations of complexity. A bird flying within a flock can position itself next to any other bird within that flock. It can change positions at will. It decides where to fly next in concert with the other birds that happen to be near it at any given time.

EMERGENCE

Complex systems arise from nonlinear interactions between their components. That's a fancy way of saying that the whole is greater than the sum of the parts. You can't capture the behavior of the flock by examining the behavior of the individual birds. The beautiful, fascinating, mysterious ebb and flow of the flock represents a property of complexity known as *emergence*. Emergent characteristics exist at the system level without any direct representation at the level of individual components.

Birds fly according to three simple rules:

- Fly in the same general direction as your immediate neighbors.
- Fly toward the same general destination as your immediate neighbors.
- Don't fly too close to your immediate neighbors.

A group of birds flying according to these rules will generate a pattern that we perceive as *flocking*. There is nothing in the rules, though, that directly explains that pattern. One might even say that the *flock* really only exists in our minds.

A complex system like a flock of birds might display emergent, coherent patterns. Those patterns, however, result from the behavior of components making individual, independent decisions. Each bird within a flock decides for itself how to respond to any given situation. A car whose doors could wander off and come back again would need a much more flexible definition of structural coherence. Otherwise it would quickly fall apart.

This difference in structural coherence illustrates a critical difference between complicated and complex systems. Complicated systems rely on centralized control and hardwired organizational structures. As a result, they work very efficiently until they break. The fact that the parts of a car all hang together is good for streamlining and thus fuel efficiency. If a wheel falls off, though, the entire car comes to a grinding halt.

By contrast, complex systems are sloppy and prone to component failure, yet highly resilient. Their decentralized, fluid structure trades efficiency for resilience. A flock of birds that encounters a giant oak tree happily splits apart and flies around it. The flock then *glues* itself back together again on the other side. A few birds might unfortunately fly into the tree. The flock as a whole, though, is unharmed by its encounter with a large obstacle. By contrast, an airplane that tried to split itself into pieces and fly around a similar obstacle would fall to the ground and crash.

Emergence presents both challenges and opportunities to organizations trying to manage complex socio-technical systems. On the one hand, it requires tolerance for failure and apparent inefficiency. On the other hand, it offers a decentralized, responsive, and scalable approach to achieving success, whether defined as control, quality, competitiveness, or profitability. Organizational methodologies that leverage the power of emergence can help companies achieve strategic coherency without sacrificing tactical flexibility.

CASCADING FAILURES

At the same time that complex systems demonstrate resilience, though, they are also subject to the phenomenon of *cascading failure*. A cascading failure is one that occurs at a higher system level than an individual component. In a complex system, failures can also result from the interactions between components.

System-level failures can even happen while all the individual parts are operating correctly.

Contemporary industrial safety research explores this phenomenon. It might be possible, for example, to extend the maintenance schedule for an airplane part without violating the acceptable wear tolerance for that part. Taken together with similar changes elsewhere within the system, however, that extension might tip the whole system into an unsafe state.

The potential for cascading failures challenges complicated-systems approaches to planning and quality assurance. Reductionist techniques that break systems into their parts are insufficient for modeling complexity. In order to understand each component and its potential to cause problems, its relationships with other components must also be considered.

SENSITIVITY TO HISTORY

Finally, complex systems exhibit what's known as *sensitivity to history*. Two similar systems with slightly different starting points might dramatically diverge from each other over time. This phenomenon is known as the *butterfly effect*. The butterfly effect describes the imagined impact that a butterfly flapping its wings has on weather patterns on the other side of the world. If the butterfly in Singapore does not fly away from a flower at precisely the time that it does, so the parable goes, a hurricane might not come into being in North Carolina.

In August 2013, the Nasdaq trading systems went offline for the better part of a day. The reasons for the outage present a fascinating example of cascading failure coupled with sensitivity to history. The "root cause" of the outage was unusually high incoming traffic from external automated trading systems. The rapidly accelerating traffic triggered a fail-safe within Nasdaq's software systems that caused them to fail over to a backup system. That same traffic level triggered a bug in the backup systems that took them completely offline.

One might point the finger at the bug as the cause of the outage. Ironically, though, the problem started because of software doing exactly what it was supposed to do: fail over based on load. That failover was intended to function as a resilience mechanism. The outage also might never have happened had the morning's traffic profile been just a little bit different. Had it not peaked quite as high as it did, or accelerated quite as quickly, the bug might not have been exposed and the failover might have worked perfectly. Or the failover logic might not have triggered at all, and the primary systems might have struggled successfully through the morning.

Emergence, cascading failure, and sensitivity to history conspire to make it infeasible to predict, model, or manage complex systems in the same ways as complicated systems. Trying to manage them too tightly using traditional top-down command-and-control techniques can backfire and turn resilient systems into brittle ones. The ability to survive, and even thrive, in the presence of failure is a hallmark of complex systems. Fires renew the health of forests. Attempts to prevent them often have the counterproductive effect of creating the conditions for catastrophic fires that destroy entire forests.

Instead, post-industrial organizations need to approach management with a newfound willingness to experiment. When prediction is infeasible, you must treat your predictions as guesses. The only way to validate guesses is through experimentation. Just as complex systems are rife with failure, so too are attempts to manage them. Experiments are as likely to return negative results as positive ones. Management for resilience requires a combination of curiosity, humility, and willingness to adapt that is unfamiliar and counterintuitive to the industrial managerial mindset.

REAL-WORLD COMPLEXITY

Complexity is more than just theoretically interesting. It increasingly presents itself in real-world business and technical scenarios. Employees have always communicated within corporations across and sometimes in flagrant disregard for formal organizational structures. In response to post-industrial challenges, businesses are trying to unleash innovation by encouraging rather than stifling complex-systems-style communication and collaboration. Management consultants are calling for the outright replacement of hierarchical corporate structures with ones that are flatter, more fluid, and more network-oriented.

The cloud is a prime example of complexity within the digital realm. A small business may run its finances using an online invoicing service from one company, an expense service from another, and a tax service from a third. Each of those companies may in turn leverage lower-level cloud services that are invisible to the end customer. If, for example, Amazon Web Services (AWS) has an outage, does the small business need to worry? The owners may not know that their invoicing service runs on top of Heroku's Platform-as-a-Service (PaaS). Even if they do, they still might not know that Heroku runs on top of AWS.

Twenty-first-century workplace trends are fundamentally changing the relationship between companies and employees. So-called Bring Your Own Device (BYOD) means that employees own their own laptops and smartphones, and can mix personal and company data on the same devices. Telecommuting and coworking move physical workspaces out of companies' control. Shadow IT lets employees buy and manage computing services without IT's control or even knowledge. Finally, companies' growing reliance on freelance labor changes the company–employee relationship at the most basic level.[2]

Together, these trends all contribute to transforming corporate environments from complicated to complex systems. They transform the management of people, devices, systems, and data from a closed hierarchy to an open network. Open organizational networks create new opportunities for business resilience, adaptability, and creativity. At the same time, though, they stress traditional management practices based on control and stability.

From Efficiency to Adaptability

Twentieth-century business structures epitomized the model of corporations as complicated systems. Companies flourished by growing in both size and structure. They mastered industrial-era technical and management practices that maximized efficiency and stability. They used these practices to create robust hierarchies that supported ever-increasing economies of scale.

Giant, long-lived companies like Ford, Johnson & Johnson, AT&T, IBM, and Kodak created and dominated entire product categories. They became known as blue-chip stocks (*http://bit.ly/1K6xhlL*) with "a reputation for quality, reliability, and the ability to operate profitably in good times and bad."

The twenty-first century features the arrival of disruption. Companies like Salesforce.com, Apple, and Tesla compete not by beating their rivals at their own game but by changing the game itself. Software-as-a-Service (SaaS) doesn't just retool software for a service economy. It challenges the very relevancy of on-premise software. Why bear the cost or burden of installing and operating your own software when you can let a service provider do it for you? Why incur large up-front capital expenditures when you can pay on-demand service fees that ebb and flow with your business?

Smartphones with high-resolution, built-in cameras challenge the relevancy of the camera as a dedicated product. Why carry around a separate photo-taking device when you can use the one that's already in your pocket? Why fumble around with SD cards to transfer your photos to a computer in order to edit and share it, when you can edit and upload directly from the device that took the pic-

---

2 Labor experts (*http://read.bi/1K6xcyi*) estimate that 40% of the U.S. workforce will consist of freelancers by the year 2020.

ture in the first place? Kodak's inability to adapt to the smartphone revolution doomed it to a rapid death after 100 years of operation.

Tesla doesn't just compete with companies that produce internal combustion-engine cars. Those companies are also producing cars with electric engines. By focusing on the zero-maintenance nature of electric cars and selling them directly to customers, Tesla is challenging the very existence of physical sales-and-service dealerships. The car dealership industry is responding not by innovating their own practices but by lobbying state legislatures to forbid direct sales of automobiles from manufacturers to customers. This tactic represents the competitive strategy of the beleaguered.

Economic transitions are fecund opportunities for disruption. As we move from industrialism to post-industrialism, new companies are disrupting incumbents by better understanding and grabbing hold of the nature of service and digital infusion. Smartphones replace cameras not just because of their physical convenience but also because of their native integration with digital services in the cloud.

Cameras were invented during the era of physics and chemistry. They represented insights into the nature of light, glass, silver, and paper. Digital photography dispenses with negatives and prints in favor of pixels. By doing so, it integrates photography with the realms of content and social media. It dissolves the boundary between pointing a physical device at an interesting building and chatting with your friends back at home about your trip to Europe. Facebook believed in the power of infusing photography and social media enough to pay $1 billion for a mobile photo uploading application.

Toyota and Chevrolet approached electric cars as a matter of replacing drivetrains. Tesla, on the other hand, took the opportunity to reimagine the entire experience of owning and operating a car. Tesla's challenge to the dealership model strikes at the heart of one of the most unsatisfying service experiences in modern life. Its cars are also deeply digitally infused. The onboard displays look more like an iPad than a traditional car dashboard. Tesla drivers can remotely open vents and lock and unlock the car using a mobile app.

In a testament to the "software is eating the world" meme (*http://on.wsj.com/1K6xNjA*), Tesla engineers its cars to work like software as much as hardware. The company provides APIs that allow third parties (including technically minded owners) to write their own applications. Tesla remotely updates its cars' onboard computer systems the way one would a website. APIs and automated updates may not sound strange or unusual in this day and age until we remember it's a car we're talking about!

FACING DISRUPTION

The pace of disruption is accelerating. Kodak was in business for over 100 years before being disrupted by digital cameras and smartphones. Microsoft went from a monopoly to an also-ran in the course of 20 years. In 2012, Nokia was the world's largest cell phone manufacturer. In 2013, when Microsoft bought it, pundits wondered whether one irrelevant company was buying another. Apple went from the world's most valuable company to a question mark—Is it being disrupted by Google? Is it being surpassed by Samsung? Has Apple lost its mojo?—in 12 months.

Disruption invokes what Clay Christensen called the *innovator's dilemma*. Disruptive innovations address customer needs left unserved by incumbents. These needs are unserved precisely because doing so would be unprofitable given the existing business model. Incumbents are often paralyzed against responding to disruptive competition because they can't get out from under their own feet.

In order to succeed in the age of disruption, companies must change their basic approach. They need to shift their emphasis from perpetuating stability to disrupting themselves. Instead of excelling at doing the same things better, faster, and cheaper, they need to challenge themselves to continually do different things, and continually do them differently. They need to learn to value learning over success, and to value the ability to change direction over the ability to maintain course. In other words, they need to shift their core competency from efficiency to adaptability.

Apple is the poster child for the growing sensitivity to disruption in the marketplace. Investors no longer judge Apple primarily by its revenues, profits, growth, or market share. They judge the company by what it's done that's new and different. Apple successfully transformed itself from a computer company to a mobile device company. Now, though, announcing a new, better, faster, bigger iPhone isn't enough. Investors and pundits clamor for a TV, a phone, or even better, something that doesn't have a name yet. The expectation isn't that Apple will improve the next device it releases, but rather that it will redefine it the way it did the cellphone.

Brands as Digital Conversations

Post-industrialism impacts companies on every level. They must truly become digital businesses. They must transform not just how they operate or organize themselves but also how they conceive of themselves. The post-industrial world-view must inform everything they do, from their most basic daily processes all the way up to the way they perceive themselves and behave as brands.

A brand represents "the unique story that consumers recall when they think of you."[3] That story reflects the promises you've made to your customers and the extent to which you've kept or broken them. These promises involve commitments to help customers and can operate on multiple levels. A sneaker company might promise to keep your feet comfortable while you run but also to help you look cool. A car company might promise to help you drive safely, but also to maintain your social status.

Companies devote tremendous energy to creating and maintaining their brands. The post-industrial economy makes brand maintenance significantly more challenging. Service transforms it from a vendor-driven activity into a conversation with the customer. Service providers must make promises about listening and responding as much as making and delivering. I judge my car company as much by its service department as by the quality of the car itself. I also judge its ability to improve its products and services over time.

Digital infusion moves the brand conversation firmly into the digital realm. I've reached the point where 99% of my relationship with my bank happens via its website. My impression of the bank's brand derives directly from the quality of its online presence. If the website is slow, clumsy looking, and hard to use, those characteristics will define the story I recall when I think of it. By making its site faster, better looking, and easier to use, the bank improves not just the quality of its online service but also the quality of its brand.

Complexity challenges companies' ability to control their brand promises. When failure is inevitable, broken promises also become inevitable. Brand maintenance thus must incorporate the ability to repair promises in addition to keeping them in the first place. The way in which companies respond to events such as security breaches become critical brand quality moments.

Companies must do more than just keep their promises. They also must make the right promises. Promising speed and handling when people value fuel

---

3 [busche2014]

economy as much as performance could degrade rather than enhance a car company's brand. Disruption complicates brand maintenance by changing the landscape under companies' feet. Tesla, for example, is disrupting the meaning of brand in the automobile industry by combining performance, luxury, and environmental sensitivity.

The post-industrial economy dissolves any remaining separation between branding and conducting business. Social media contributes to this trend by wresting control of a company's brand away from it in favor of consumers. No longer does the company drive the public's impression of it. Instead, companies become mere participants in the discourse about their identity, quality, and value.

Post-industrialism turns brand management into a digital conversation between a company and its customers. Brand quality depends on a company's ability to conduct that conversation as seamlessly and empathically as possible. Having lost control of the message, companies must shift their focus from trying to shape their customers' perceptions, needs, and desires, to accurately understanding and responding to them. The capacity for empathic digital conversation thus becomes the defining characteristic of post-industrial business. The ability to power the digital brand conversation in turn becomes the defining measure of quality for post-industrial IT.

The New Business Imperative

In order to shift their approach to brand management from a broadcast model to a conversational model, twenty-first-century businesses must simultaneously transform the way they relate with their customers and the way they organize their internal operations. Conversation depends on the ability to listen and to respond appropriately to what you've heard. Digital businesses thus must organize themselves to accurately, continuously process market and customer feedback.

The ability to process feedback fluidly is a critical component of post-industrial business success. Service requires conversational marketing and co-creative business operations. Infusion requires deep integration between technical and business concerns across physical and virtual dimensions. Complexity requires exploration and tolerance for failure. Disruption requires an unfettered ability to uncover and pursue new possibilities.

These requirements necessitate an internal transformation that mirrors the external one. In order to provide high-quality, digitally infused service, the entire delivery organization must function as an integrated whole. In order to handle the perturbations caused by complexity and disruption, that same organization must be able to flex and adapt. The post-industrial company thus begins to look more like an organism and less like a machine.

Post-industrial businesses are beginning to experiment with nontraditional organizational structures. Dave Gray has described networks of pods in his book, *The Connected Company*. Zappos, an online shoe retailer, is experimenting with holacratic management structures that distribute decision-making throughout self-organizing teams. Amazon CEO Jeff Bezos coined the phrase "two-pizza team" to refer to small, integrated teams that have the autonomy and intimacy necessary to move quickly (and can be fed by just two pizzas). Yammer, an enterprise collaboration software vendor purchased by Microsoft in 2012, dynamically creates similarly sized functional teams. Each team dissolves and reorganizes after every project in order to propagate knowledge throughout the larger organization.

These new structures leverage the power of emergence to balance flexibility with coherency. In order to use them successfully, however, twenty-first-century businesses need more than a new org chart. They need a new worldview from which to operate, one that shifts the emphasis of management values, goals, and structures, and the IT systems that support them:

- From efficiency, scale, and stability to speed, flexibility, and nimbleness
- From discontinuous broadcast to continuous conversation
- From avoiding failure to absorbing it
- From success as accomplishment to success as learning

2

A New Model of Control

Post-industrialism challenges twenty-first-century businesses to become more open, responsive, experimental, and resilient. Companies still, though, exist to generate profit. They aren't indifferent to the results of their efforts. Profit isn't something companies are happy just to let ebb and flow naturally. They direct all their efforts toward controlling it and ensuring that it continually moves in the right direction.

Profit allows a company to perpetuate its existence, and to replenish and grow itself. Growth lets public companies satisfy the imperative to create shareholder value. Even small, private companies pursue profit and growth in order to generate financial fuel for personal goals like sending your children to college or affording retirement.

One might say that control is the most basic activity that defines a corporation. Much of its activity will be directed toward controlling the parameters that contribute to profit and growth. In the quest to maximize profit, companies must control both internal and external parameters. Internal parameters include cost of materials, lead time, product quality, information flow, operational procedures, and employee behavior. External parameters include market share, income, customer behavior, and stock price.

IT came into being to aid the quest for control. Traditionally it provided information needed by internal and external control mechanisms. As companies have increasingly automated internal processes and customer interactions, IT has become part of the control mechanism itself. Understanding how post-industrialism changes the nature of control is critical to understanding IT's new mandate.

The Industrial Model of Control

As much as anything, the industrial era will be remembered for the emergence of powerful business control mechanisms. The industrial model of control is largely unidirectional. Information flows down and out along hierarchical management chains. Production flows forward through assembly lines and out to customers. Marketing (the attempt to control customer behavior) flows out from design agencies to media outlets to consumers.

Industrial control systems focus on maximizing efficiency in order to optimize unidirectional flow. Scientific management techniques such as Taylorism and Fordism drew inspiration from time-and-motion studies. They looked for ways to wring inefficiencies out of the production process. This approach implied that it was possible to analyze a process, identify and remove waste, and then operate the improved version in an ongoing, steady-state fashion. Having identified a more efficient operating model, one could implement that model and then run it unchanged for lengthy periods of time.

Frederick Winslow Taylor

Frederick Winslow Taylor was a mechanical engineer who sought to improve efficiency and working conditions in industrial factories by applying scientific principles to their management. He used experimentation and observation to identify repeatable, optimally efficient, individually manageable tasks. Taylor's model relied on replacing individual techniques for accomplishing tasks with standardized methods. Management becomes a matter of training workers in these methods and overseeing their work to make sure they adhere to them.

Taylor radically separated the definition and management of work from its execution. That separation underlaid Ford's approach to automobile production and to some degree the whole industrial model of business management. It had the advantage of maximizing control and efficiency for complicated activities. It had the disadvantage of creating brittle processes that limited workers' ability to adapt to unforeseen circumstances. It also relied on a primitive understanding of worker motivation that inhibited its efforts to improve workplace productivity and satisfaction.

Separating design from implementation and management from execution was central to Taylorism and Fordism. Industrial control relies on decomposing global mechanisms into subcomponents and removing variation within them. On a human level, industrial control mechanisms view workers as integral parts of an overarching machine. The production process has primacy over the people participating in it. Designers develop an overall control model, and then break it down into component parts. Managers instruct and oversee workers to ensure work happens according to plan.

The Legacy of Henry Ford

Henry Ford didn't actually invent the automobile. What he did invent was a repeatable, efficient, scalable manufacturing process. His dictum about the Model T that "you can get it in any color you want, as long as it's black" reflected his strategy of restricting choices in order to simplify assembly line requirements. That strategy allowed Ford to dramatically reduce production costs. By doing so, he revolutionized society by making automobiles affordable for the masses.

Ford's solution to mass production epitomized the industrial approach to value and its means of creation. It focused on quality through similarity. Everyone who bought a Model T knew what they were getting. They could rely on its reliability thanks to the carefully controlled production process. The assembly line achieved quality by reducing variety, both in terms of what was being made and how it was produced. Not only did Ford dictate that every car would be black; he also dictated what every worker would do at every point in the production process.

Ford's innovations profoundly influenced the course of twentieth-century business. Companies treated manufacturing as a process of squeezing variety and inefficiency out of the value stream. They treated management as a process of ensuring worker compliance with that process. Mad Men–style marketing departments evolved to convince customers to "consume" the standardized set of products that flowed outward from them.

Corporate IT arose as a tool for managing industrial-style control systems. It reflected the push model of product companies. IT helped companies track business-control parameters such as inventory, cost, sales, and defect rates. Workflow management systems helped manage procedures that flowed through complicated-system organizational and process structures. Manufacturing automation systems increased efficiency and quality on industrial shop floors. IT even developed industrial-era control models such as ITIL[1] to manage itself.

The Limits of Industrial Control

Post-industrialism is forcing companies to rethink their reliance on industrial control mechanisms. It challenges the effectiveness of designing static, long-running production processes. It questions whether optimization is feasible or even desirable. It confounds organizational structures based on hierarchy and unidirectional information flow. It eludes the capabilities of workforces that are at once conformist and disconnected.

Service, digital infusion, complexity, and disruption all contribute to challenging the applicability of industrial control mechanisms to the post-industrial economy. Service necessitates conversation and collaboration between company and customer. Controlling the customer relationship becomes a matter of satisfying rather than convincing. In order to satisfy customers, a service provider must be able to listen. The entire organization must be internally aligned so it can hear the customer accurately and then cooperate to generate an appropriate response.

Compartmentalized, hierarchical corporate organizational structures, characterized by one-way information flow, impede the ability to listen, hear, and respond. This impedance mismatch shows itself through all-too-familiar customer satisfaction failures. We've all experienced the frustration that comes from asking a company representative for help only to be told "that's not my department." That response might be appropriate from the point of view of optimizing internal efficiency. From a global control perspective, however, it has backfired. By frustrating customers, the interaction incentivizes them to find an alternative service provider and take their money elsewhere.

Digital infusion further stresses compartmentalization. No longer are systems of record disjointed from systems of engagement. Not only must the systems interconnect; their design, development, and operational processes, and the people who run them, also must interoperate. IT organizations generally manage systems of record using rigid processes that intentionally slow down the rate of

---

1 *http://en.wikipedia.org/wiki/ITIL*

change in response to constraints such as regulatory rules. Systems of engagement, on the other hand, must respond to marketing pressure for quick and continual change. The digitally infused business must find a way to blend or at least synchronize these differing approaches.

COMPLICATED CONTROL OF COMPLEX SYSTEMS

The industrial company is the epitome of a complicated system. It thrives on predictability and stability. Industrial systems management views failure as something to prevent. It tries to create fail-safe systems by structuring components into rigid layers. Each layer depends on the robustness of the layer below it. Robustness within a layer becomes a matter of trying to maximize the mean time between failures (MTBF) of each component.

Top-down industrial control poses the greatest threat to complex systems such as post-industrial organizations and IT systems. These systems excel at responding to failure, but wither in the presence of too much stability. Fail-safe design solutions create brittleness instead of robustness by thwarting the open, dynamic connections that characterize complexity. Due to their sloppy, failure-prone yet resilient nature, complex systems need the ability to absorb component-level failures instead of trying to prevent them. They also need the ability to continuously recalibrate their survival strategy in the face of ever-changing environments.

DISRUPTION AND INDUSTRIAL CONTROL

The assembly line is the central image of industrialism. Even nonmanufacturing companies use the assembly line as a design metaphor for business processes. In the age of disruption, the assembly line is problematic. It relies on significant up-front investment in design and optimization. It assumes the ability to develop a stable production mechanism, which can then be operated without major changes for long periods of time. It makes rapid change difficult. As a result, it amplifies the Innovator's Dilemma by incentivizing companies to keep doing what they know how to do in the ways that they know how to do it.

Twenty-first-century businesses still need to generate continuous profit and growth. Now, though, they need a new way to drive the economic engine. They need the ability to synchronize themselves with their customers and the market much more intimately. When brands become conversations, the problem of controlling your brand requires a new strategy.

To accomplish the level of synchronization that brand conversations need, post-industrial businesses must minimize the delay between incoming signals and outgoing replies. They need the ability to continually change themselves in response to those signals. They need a mechanism that allows them to navigate uncertain, unknowable, continually changing environments. In other words, they need to turn the familiar, industrial model of control completely on its head. Because IT is integral to controlling the digitally infused business, it must fully participate in this transformation.

Cybernetics: A Post-Industrial Model of Control

Ironically, in the search for a post-industrial model of control, we can look all the way back to the dawn of computing in the mid-twentieth century. The work of Norbert Wiener and others in the 1940s and 1950s offers a powerful model for adaptive *control via conversation*. Wiener was a major contributor to the original conceptions of computing and information theory. In the process, he led the creation of a new, holistic, cross-functional discipline called *cybernetics*.

During World War II, Wiener helped the US military with its attempts to develop automatic targeting for anti-aircraft guns. It was his idea to use knowledge about physics, airplane design, and pilots' cognitive processes to predict an attacking plane's flight path. Wiener recognized that a plane couldn't instantaneously change its position from one random position in the sky to another. If it did, it would tear itself to pieces. Its evasive capacity was also influenced by limitations on the pilot's ability to calculate new maneuvers. The more quickly he had to respond, the more likely he would be to resort to habitual maneuvers. Wiener believed it should be possible to mathematically compute probable flight paths, and use those computations to automatically point the gun at a location that was likely to intersect with the plane's path.

Wiener's research confronted him with two problems. First, it wasn't possible to perfectly predict the path of a human-guided physical object through physical space. Second, even if that were feasible, it still wasn't possible to perfectly aim a large, heavy gun barrel. Depending on environmental parameters such as air temperature, humidity, and even the age of the grease in the gun turret's ball bearings, the barrel might swing a little too far or not quite far enough.

Wiener compensated for these unavoidable inaccuracies by building a feedback mechanism into the targeting system. It fed information about the gap between the intended and actual aim, and the predicted and actual flight path, back into the targeting system. Rather than just guessing the plane's future location as best it could and then pointing and firing as best it could, the system successively approximated the desired aim, correcting itself along the way.

After the war, Wiener collaborated with Arturo Rosenblueth, a researcher in physiology at Harvard. Together they explored various biological functions, including one known as proprioception. The human body uses proprioception to control physiological processes such as walking. It loosely refers to *muscle sense*, or feedback from muscles, which allows the brain to judge progressive movement toward a target. A field sobriety test, for example, tests for impairment in your ability to successively move your finger toward your nose.

When you walk, your brain evaluates information telling it how far your foot still is from the ground. It then moves your foot closer, and reevaluates the distance to the ground until your body makes the next step. This process comprises a continuous, feedback-based control loop. The brain directs the leg to act, then listens to information "fed back" to it about the result of its action, then directs the leg to act some more, and so on. This process is illustrated in Figure 2-1.

Figure 2-1. Basic feedback loop

Wiener's work on weapon targeting, adaptive behavior, information theory, and proprioception all contributed to the development of cybernetics. Cybernetics formalized the science of feedback. The name comes from the Greek word for *steersman*. It shares its etymology with the word *governor*.

Cybernetics treats control as a dynamic process of maintaining homeostasis by continually processing and responding to feedback. It uses information about outcomes in the past to control actions in the future. If the gun barrel swung too far, by how much was it off? How inaccurate was our prediction of the plane's next evasive maneuver? How far do I still need to move my foot to reach the ground?

Circular causality is a key characteristic of cybernetic control. Feedback loops push information about behavior from the past forward into future decisions. Cybernetic processes such as proprioception define circular relationships between the brain and the body. We generally think about the brain–body relationship in industrial terms, with the brain as manager and the leg as worker. According to proprioception, though, the brain paradoxically depends on the leg to inform its decisions.

Cybernetics assumes a constantly changing, unpredictable world through which we must steer our way. We can't perfectly know either the present or future state of our environment. We must therefore incorporate listening into our actions so that they become inseparable parts of a continuous process. A sailor or oarsman, for example, must continually compensate for changes in wind or current. Without going in a perfectly straight line, the boat nonetheless wends its way under the steersman's guidance to the desired destination on the other side of the lake.

Cybernetics provides a language that expresses the essence of post-industrial business. In addition, that language describes the fundamental nature of post-industrial IT. Contemporary methodologies such as Agile and DevOps, for example, use feedback loops to let development and operations teams steer in response to continually changing a requirements.

Second-Order Cybernetics

Wiener and other early cyberneticians participated in a series of symposiums in the 1940s and 1950s called the Macy Conferences. These conferences brought together researchers from a wide variety of disciplines, including mathematics, biology, psychiatry, and sociology. Together, they hatched the beginnings of systems thinking, complex systems studies, and cognitive science. They explored the potential applications of cybernetics to everything from modeling the human mind, to rethinking psychological counseling, to managing businesses, to steering entire national economies.

The first-generation cyberneticists tended to take a mechanistic approach to controlling systems of all kinds. Wiener introduced the new discipline to the world in 1948 with the publication of his book, *Cybernetics: Or Control and Communication in the Animal and the Machine*. Much of the original cybernetics work thought more in terms of control and machines than communication and animals.

This approach could become problematic when applied to complex systems such as people and societies. On the one hand, it does makes some sense to think about mental health as operating within a desired range of behavior. The psychologist could be seen to provide a feedback-based regulatory function, helping the patient bring themselves out of neurosis and back into homeostasis. On the other hand, an acceptable doctor–patient relationship must respect the intelligence and autonomy of both parties. The therapeutic relationship itself becomes ill if it views the patient as a device to be controlled.

Very quickly though, some of the original cyberneticians began thinking at a higher level. Margaret Mead and Heinz von Foerster introduced the concept of second-order cybernetics, also called the cybernetics of cybernetics, or *New Cybernetics*. It distinguished, in von Foerster's words, between "the cybernetics of observed systems" and "the cybernetics of observing systems."

Second-order cybernetics is inherently holistic. It recognizes that a first-order cybernetic system such as a thermostat is itself cybernetically controlled, in this case by the human who programs it. A thermostat has no purpose without the room that contains it, along with the air it's controlling. The room in turn exists within a house. The person who lives in that house sets the thermostat so that it's warm in the morning when they get up to go to work. They go to work because they're employed by a company. They drive to work in a car that was made by a car manufacturer, on a road that's maintained by the town in which they live.

On another level, second-order cybernetics restates relationships such as therapy in terms of interactions between autonomous systems that are themselves complex. It emphasizes the animal and communication over the machine and control. It presents a model that is interactive and mutual rather than manipulative and hierarchical.

Von Foerster's concept of "the cybernetics of observing systems" captures the fact that complex systems always interact with one another from their own points of view. None of them holds objective knowledge. Instead, they negotiate shared understanding through conversation. That shared understanding is also doomed to imperfection. It continually breaks down and must be repaired as part of the dynamic process of life.

CYBERNETICS AND POST-INDUSTRIAL BUSINESS

Cybernetics views existence as a co-creative conversation with your world, rather than a disconnected process of ingesting and interpreting information from it. Though it might sound like something out of postmodern philosophy, this worldview is directly relevant to post-industrial business. Digitally infused service breaks down traditional boundaries between vendor and customer, between virtual and physical, and between internal and external systems and processes. In order to provide quality service in such an environment, we need to augment our traditional, industrial-era reductionist thinking with holistic, integrative systems thinking.

In order to change our approach to value from delivery to co-creation, and our approach to brand from broadcast to conversation, we need to embrace the cybernetic view of reality as conversationally co-constructed. Business no longer consists of making things and then trying to convince people that these things are what you claim they are or useful in the ways you claim. Particularly in the era of social media, things are what people perceive them to be. The nature of a service is inseparable from its use; its usefulness arises from the collaboration between vendor and customer.

Complexity defies objective knowledge. Not only can we not fully model a complex system at any point in time and not only does the system we're trying to model continually change, but our attempts to model it contribute to changing it. Rather than managing complex systems from above, we engage with them as participants. Our understanding of the systems we create, manage, and operate reflects our particular perspective as observers. It is necessarily relative and provisional.

Given that complexity increasingly characterizes our businesses and the IT systems we use to manage them, we need a new model of control. Cybernetics provides a model that accurately reflects the interactive nature of our relationships with these systems. That model frees us from the temptation to try to harness complexity through industrial-style control. It allows us to understand the limits of our ability to predict complex behavior, and a concrete method for replacing our attempts at prediction-based, top-down control with feedback-based steering.

Autopoiesis: Self-Steering Through Conversation

Cybernetics' insights into the co-creation and circular causality found their fullest expression in the work of Humberto Maturana. Maturana is a Chilean biologist who studied with Warren McCulloch, von Foerster, and other founding cybernetics luminaries. In particular, it was the biological experiments that he helped McCulloch conduct that set Maturana on the path to his unique contribution to biological cybernetics.

Maturana and McCulloch were investigating the visual systems of frogs. They made a surprising discovery: a frog's vision is specifically constructed to see small, fast-moving objects (such as flies), and to ignore large, slow-moving objects. In other words, the frog sees the world in a particular way not because of how it interprets reality, but because of its internal structure.

These experimental results led Maturana to reconsider his views on reality, knowledge, and cognition. Being a biologist, he found himself struggling with the essential definition of what it means to be a living system. He arrived at a definition of *life* that, in typical cybernetic fashion, required the invention of a new word derived from Greek. In his case, it was *autopoiesis*.

Autopoiesis means *self-production*. Maturana distinguished it from *allopoiesis*, or *other-creation*. A thermostat exists in order to change the temperature of the air in a room. A word processor exists in order to produce text documents. A living system, on the other hand, exists in order to perpetuate its own existence.

Autopoiesis is the process by which components of a system work together to create the conditions for their own production (the process is illustrated in Figure 2-2). A human being, for example, exists through the interaction of lungs, heart, brain, and muscles. By continuously finding food and shelter, the human makes it possible for the lungs, heart, and brain to continue to function. If I develop lung cancer, I need to get treatment for it. If I don't and my lungs fail, I will die. I die not just because I have lungs but because they provide a function upon which the other parts of me (as well as me as a holistic living system) depend. Without my lungs, the rest of my organs can no longer work together to keep me alive.

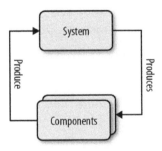

Figure 2-2. Autopoiesis

Autopoiesis may seem like a solipsistic view of life, but it only represents part of the story. Maturana explored autopoietic systems and their relationships to their environments in a series of books he coauthored with his student and colleague, Francisco Varela. According to them, a system's autopoiesis must be appropriate to its environment in order to be sustainable. They called this requirement *structural coupling*. If a frog is constructed to see flies, for example, but there are no flies in the frog's environment, then it will die.

Structural coupling happens by way of perturbation. Living systems induce behaviors within one another as part of sharing an environment. Frogs jump and stick out their tongues because they see flies. Antelope look around and flick their tails because they share water holes with lions. People sleep in tents with netting because they live in environments with mosquito-borne malaria.

Structural coupling is not, however, a static process. Autopoiesis always occurs within an environment that is continuously changing and evolving. In order to maintain its structural coupling with its environment and thus its viability, a living system must also change and evolve. Maturana and Varela used the term *self-steering* to refer to the process of autopoietic adaptation.

The concept of self-steering posits that living systems use a cybernetic process to maintain their autonomy within their environment. An individual's response to a lung cancer diagnosis is such a mechanism. By taking a leave from work to receive chemotherapy treatments, that person is changing his situation in order to maintain autonomy—that is, to stay alive. Self-steering paradoxically lets a living system change, or adapt, in order to stay the same.

Self-steering implies that living systems don't operate in slaved response to stimuli from the outside world. Their structural coupling with their environment causes them to respond to outside perturbations through adaptation. In the process they perturb their environment, causing the other autopoietic systems in that environment to adapt in turn. Autopoietic systems *swim around*, as it were, co-evolving along with other living systems, jointly creating a shared, dynamic reality.

Maturana and Varela introduced the theory of autopoiesis in the book *Autopoiesis and Cognition*. Their use of the word *cognition* refers to the claim that cognition is nothing more nor less than the mechanism by which a living system (re)-creates itself through self-steering. Cognition is therefore common to all autonomous systems. According to Maturana and Varela, the presence or lack of an advanced nervous system is irrelevant to the ability to cognize. Companies and the organizational subsystems (teams, departments, pods, etc.) that compose them can self-steer in the same manner as cells, fish, or people.

Cybernetics concerns itself with adaptive, purposeful systems. The purpose of an automatic gun turret is to shoot down planes. The purpose of a thermostat is to keep the temperature in a room comfortable. Maturana and Varela's special contribution to systems thinking was their insight that a living system is not just a static arrangement of interrelated things. Its structure is a continuous, dynamic process of cognition for the purpose of staying alive. Because its surrounding environment is continually changing, the system must continually change itself in order to fulfill that purpose.

The Self-Steering Organization

You may be wondering about the relevancy of Maturana's and Varela's work. It sounds abstract and esoteric, and revolves around a word that's nearly impossible to spell! To appreciate its applicability, you need only consider the nature of business. What is the purpose of a company? On one level, a company might have a noble, or perhaps just a cynical, goal within its community. On a more primal level, though, if the company doesn't stay in business, it can't achieve any of those larger goals.

A company exists as a network of employees, IT systems, physical plant, teams, departments, and divisions, all of which conspire together to create products and services that the company can sell. Gaining revenue through sales allows the company to pay for its employees and IT systems and physical plant in order to create and sell more products and services.

One could thus view a company as an autopoietic living system. Furthermore, one might claim that the post-industrial economy—characterized as it is by service conversations, complexity, and disruption—requires a self-consciously autopoietic business strategy. When continual change becomes the primary characteristic of your environment, continual internal and external adaption must become the primary capability of your operating model. The key success criterion of IT in turn becomes its ability to power a business's autopoiesis.

The history of Apple offers a perfect example of the corporation as an autopoietic system. During his introduction of the Apple Watch, CEO Tim Cook repeatedly used the word *personal*. Many readers may not be old enough to fully appreciate this reference. In the 1980s, Apple shipped the Apple IIe in a box with the slogan "The Personal Computer" printed on the side. The idea was that the IIe and the Macintosh after it were very personal devices. The IBM PC, by contrast, was (at least in Apple's mind) a very impersonal device.

Since then, Apple has undergone dramatic changes. They've gone through three profound CEO transitions: from Steve Jobs to no Jobs, from no Jobs back to Jobs, and from Jobs to Cook. The company has hired and fired people, restructured its org chart, and built new buildings. Recently, it's gone so far as to recruit fashion industry executives and Australian furniture designers. It's shifted from being a computer company to, according to some analysts, primarily being a telephone handset company. It's on the way to becoming—who knows? A music industry company? A banking company? A jewelry company?

Through all of these changes, though, Apple has stayed in the business of creating *personal* experiences. It is the quintessential self-steering organization.

On the one hand, it's changed in nearly every way. On the other hand, it's undergone these changes precisely for the purpose of maintaining its essential identity in the face of a profoundly transformed environment.

The Cybernetic Insight

Maturana's definition of living systems as autopoietic continued the tradition of cybernetics's fascination with circular causation. A living system embodies a circular relationship between the system and its components. Maturana's biological background led him to define that relationship in terms of processes. Components participate in the operation of a holistic process that in turn creates the components. Life requires continual action; stasis results in death. The dynamic nature of viable systems applies equally to cells, people, businesses, and societies. In each case, what they are is what they do and how they change in response to their environment.

The theory of autopoietic self-steering carried on the simple yet profound conception of reality as a continual unfolding that marks all of cybernetics. Although first-generation cybernetics tended to think in terms of control, its understanding of the nature of control was revolutionary. The most basic definition of cybernetics contains Ranulph Glanville's observation that "the controller is controlled by the controlled." Every cybernetic process implicitly involves conversation and holistically points upward toward systems and away from reductionism.

The thermostat, for example, is the "Hello, World" of cybernetics. The truth is that thermostats don't really *control* the temperature of the air in a room. It's not as if a thermostat can force the air molecules to move at a specified average speed for some amount of time. Instead, it *listens* to the current temperature (by being physically deformed by the air around it). Based on what it *hears*, the thermostat influences the air to change its temperature by introducing warmer air from a furnace. The air in the room happily lets itself be influenced, but thanks to the second law of thermodynamics, refuses to behave and quickly starts cooling off again. The thermostat is thus forced to continually adjust. It is the controlled as much as the controller.

By considering Maturana's work on autopoiesis, autonomy, and self-steering, along with revisiting the nature of thermostats, we can restate cybernetics from a less mechanistic, control-oriented perspective. Cybernetic systems interact with one another by way of conversations. In order to maintain existence, autonomy, and integrity, one must be able to listen. Autopoietic systems must listen to one another and to themselves (or more accurately to their component parts). Living systems co-create reality through circular influence. Circular influence is, and happens via, a conversation.

Cybernetics as a Model for Post-Industrial Control

Cybernetics was extremely popular and influential in the 1950s and 1960s. For various reasons (some having to due with Norbert Wiener's eccentric personality), it began to lose influence and has largely been forgotten. Its fate is ironic, given the popularization of innumerable words that start with the prefix *cyber*.

Regardless of its history, cybernetics is more relevant than ever. By interleaving planning and implementation into a unified process, cybernetic systems transform execution into continuous learning. This kind of *control as conversation* is exactly what post-industrial businesses need to confront the challenges posed by service, infusion, complexity, and disruption. In fact, we can see the trace of cybernetics in many of the methodologies companies are adopting to confront twenty-first-century business imperatives.

LEAN STARTUP: CYBERNETICS'S SPIRITUAL INHERITOR

Lean Startup is a popular new product development methodology that exemplifies the cybernetic model. Lean Startup warns against committing up front to large-scale, long-term plans for implementing a product vision. If we do commit too early, we risk wasting resources building something that doesn't match the market and therefore doesn't deliver value either to the company or the customer.

Instead, Lean Startup counsels us to build the smallest possible version of our vision that lets us test it against the market as quickly as possible. Feedback from our tests provides insight with which to refine our vision and our implementation. By executing this build-test-learn cycle repeatedly, we steer our way toward a truly viable product.

Lean Startup incorporates the idea of a *pivot*. Sometimes what we learn is that we're trying to solve the wrong problem. We might have misunderstood the problem in the first place or it may have changed. In either case, we need to pivot and start learning/steering in a new direction.

If this all sounds familiar, it should. Imagine the customer as a plane you're trying to capture. Your product vision is your prediction of the customer's flight path: if I build something with functionality X, it will match up with the customer's needs at point Y. The build-test-learn cycle reflects the recognition that you can't perfectly predict the precise location of point Y, nor can you perfectly aim your product there. You need to incorporate feedback in order to successively correct your aim. *Pivoting* happens when you learn that you mis-predicted customers' goals and desires (or that they changed them in the midst of your product development).

First-order cybernetics helps us understand how to manage individual activities such as product development projects. Second-order cybernetics offers a model for thinking about businesses as wholes. We can view companies and their customers as autopoietic, self-steering systems. Companies exist for the purpose of maintaining their own existence (sustained negative cash flow leads to a company's death, whereas profit allows it to replenish and grow its components and thus itself).

Customers, for their part, need to interact with companies as part of their own autopoiesis (without someone to deliver heating oil, I can't use my thermostat to keep my house warm). Companies self-steer in the face of threats to their autonomy: a company introduces a new product to ward off a competitor. Customers do the same: if heating oil is too expensive for me to afford, I try to find a less expensive supplier or switch my heating system to use natural gas.

Post-industrial businesses need to replace the assembly line with the thermostat, the turret gun, and the living organism as their organizing metaphors. No longer can they design their systems and processes for linear efficiency. No longer can they afford to devote significant time to up-front development under the assumption they can run a single "assembly line" at length once they've optimized it. Instead, they must design and evaluate systems and processes for their ability to process and respond to feedback: to continuously recognize and measure inaccuracies in their customer understanding, and to recalibrate themselves based on those measurements.

In the age of disruption, efficiency becomes a measure of how quickly and accurately businesses and their component parts can adapt. Companies' ability to survive as autopoietic systems depends on their ability to self-steer through conversation and the effectiveness with which their IT organizations can power that conversation. Even the simple thermostat can quickly and easily be reprogrammed at the whim of the home's occupant. In the post-industrial economy, the quality of the thing gives way to the quality of the conversation.

Cybernetics as a Unifying Perspective

In order to help customers accomplish their goals, digital service businesses need a unity of purpose across the entire organization. At the same time, complexity and disruption necessitate the ability to flex and change internally. Autopoiesis, expressed as self-steering through conversation, provides a unifying perspective that lets companies resolve these apparently contradictory needs. The metaphor of self-steering captures the mechanism by which organically organized, complex adaptive businesses interact with their environments.

Autopoietic systems function through circulatory causality. Their components exist because of and in service of the system. Lungs have no purpose without a body and can't survive outside of it. At the same time, the system exists because of and in service of its components. The body contains lymphatic and circulatory systems that keep the lungs working. Without all of those subsystems, there is no body. The body also comprises eyes and ears that let it sense the environment. When the ears hear a lion and the legs make the body run away, the eyes and lungs (along with all the other body parts) survive to face another day.

We can think of a post-industrial company in similar terms. Teams and departments have cybernetic, conversational relationships with each other. They are structurally coupled with each other and adapt to each other's perturbations. The corporation as a whole, and its vision and mission, define the environment within which those conversations take place.

The company simultaneously carries on a conversation with its customers by way of its components. Marketing, finance, product development, and IT all must converse with one another and with the customer. Based on feedback from customers, those teams all contribute to helping the corporation adapt and survive through its own structural coupling with the competitive market.

Just as a biological system self-steers by way of mutual service between system and components, so too does a corporation. The company co-creates value with customers through service. In order to make that exchange of value possible, the parts of the organization (marketing, design, development, operations, finance, etc.) all need to co-create service value with one another. Service at all levels is mutual and conversational, and thus cybernetic.

Cybernetics and Empathy

Successful self-steering depends on empathy. Feedback represents an outside perspective. In order to properly respond to feedback and thus steer productively, an organization must accurately understand that perspective on its own terms.

It's important to understand that empathy does not mean understanding someone else's needs, nor does it mean feeling sorry for them. Empathy is rather "the intellectual identification with or vicarious experiencing of the feelings, thoughts, or attitudes of another." The ability to empathize is the ability to understand a situation from another's point of view.

This definition doesn't imply wallowing in another's pain. Just because we can empathize with a depressed friend doesn't mean we become similarly paralyzed by gloom. Although explanations of empathy often use painful experiences to differentiate it from sympathy, this definition doesn't necessarily imply a relation to pain at all. For example, it's possible to empathize with other people's aspirations as well as their frustrations.

Empathy can also function on a much more mundane level. Why does someone navigate a user interface the way they do when there are "more efficient" ways to do it? Why do they take a certain route to work when there's a shorter way?

Empathy is the basis for systems thinking. Systems emerge from relationships between components. Relationships happen through conversation. Conversation requires the ability to comprehend what another party "is trying to say to you"; in other words, the ability to see things from the other's perspective. Otherwise, the participants in a relationship are just talking at each other without actually relating to each other. Without relationship, there is no system.

A self-steering system maintains structural coupling by allowing itself to be perturbed by its environment. In a sense, it "lets its counterpart in." By seeing itself from another's point of view, a living system can steer toward continued life and away from impending death. This impetus applies just as much to businesses relating to their customers and competitors as it does to frogs and human beings. The purpose of a business might be *selfish*; without the ability to see itself from the perspective of the market, however, a company will lose customers, and thus revenue, and thus viability.

According to designer and researcher Seung Chan Lim's book *Realizing Empathy* (*http://amzn.to/1K6CJoN*), true innovation requires empathic conversation. Without it, *innovation* remains merely a generator of novelties that are interesting without being useful. A successful business doesn't just need the ability to change; it needs the ability to change in the right direction, at the right time, for the right reason. Empathy makes change meaningful rather than random.

Lim describes the development of empathy as an unfolding process of speaking, listening, and processing. The participants in an empathic conversation progressively uncover each other's perspectives. In other words, they use feedback loops to steer their way toward mutual recognition: "No, that's not what I meant; what I really meant was this...". One could say that cybernetics generates empathy, and that empathy is cybernetic.

Cybernetics's approach to control is so radical because it incorporates listening and responding. The word *conversation* comes from the Latin for "to turn with." In order to converse with someone, you have to continually switch back and forth between speaking and listening. A feedback loop cycles between acting and asking. Successful control, in the form of accurately understanding and responding to the answer that comes back from the environment, requires empathy. As Heinz von Foerster succinctly said, "I like cybernetics: its intrinsic circularity helps me see myself through the eyes of the other."

IT as a Cybernetic Medium

While people often think of cyborgs when they hear the word "cybernetics," nothing about it necessarily implies artificial intelligence or even computerization. Cybernetics is a conceptual model for the way various kinds of complex systems relate to the world. Particularly in Britain, cybernetics evolved as a way of thinking about human interaction and organizational management. Stafford Beer, a leading British figure in the field of operations research, applied cybernetic principles to the management of companies and even entire economies.[2]

Modern businesses do, however, rely on computerized systems as a critical aid to managing themselves. IT both reflects and makes possible certain kinds of information flow, activities, and relationships within organizations. Business operations and IT mirror each other. A business that seeks to manage itself and its customer relationships cybernetically needs an IT organization that takes the same approach to its systems, practices, and relationship to the larger business.

Digital infusion makes the relationship between IT and the surrounding organization even more intimate. IT already has become the mechanism through which companies operate themselves and communicate internally; now it needs to extend itself to become an enabler for empathic customer engagement. In the digital service economy, companies' ability to co-create value with customers, as well as their ability to sustain themselves through adaptation, both depend on IT.

---

2 [medina2014]

Post-industrial companies need a medium through which they can conduct the digital conversations that define their brands. This medium must be cybernetic in nature. It must enable self-steering, allowing companies to respond fluidly to perturbations from their customers and from the market. It must also allow the components of those companies to respond to perturbations from each other. In order for IT to take on that role and become a digital conversational medium, it will have to deeply absorb the cybernetic perspective on systems and relationships.

3

IT as Conversational Medium

Serving as a medium for digital conversation is IT's new mandate. Responding to this mandate means radically transforming IT's understanding of its purpose, as well as its approach to fulfilling that purpose. It means replacing information systems management with empathic conversation as both the driving force and the ultimate goal by which post-industrial IT measures itself.

IT has traditionally taken an industrial approach to building and operating systems. It has concentrated on implementing large-scale solutions with high up-front costs and an equally large presumed return on investment. Development has proceeded in linear phases, from initial conception, through design, development and testing, to completion. Project management has used sophisticated tools and processes that have required specialized expertise.

IT took this approach under the impression that it could, and must, predict needs and capabilities in advance. Business requirements were seen to be highly complicated and thus to need highly complicated solutions and solution implementations. IT organizations invested millions of dollars in software solutions such as ERP systems that were intended to model the entire business.

All-encompassing enterprise projects took years to deploy and often cost as much to implement as they did to buy. Sadly, more projects failed than succeeded. IT developed a reputation for an inability to reliably deliver on its claims. Large-scale IT projects generally failed for three reasons:

- The project plan failed to accurately predict or account for the challenges that inevitably accompany complicated system implementations
- The requirements failed to accurately model users' real needs

- Over the course of a lengthy implementation project, users' needs changed

From a cybernetic perspective, the reasons for failure are obvious. Traditional IT project planning and management had no mechanism for incorporating feedback. It had no way to self-correct other than by throwing away months or years of work and starting over.

IT's inability to reliably or efficiently deploy useful solutions is just the beginning of its troubles. It also struggles to operate those systems. Once an IT organization has managed to deploy a complicated, expensive software system, it needs to keep that system running. Change introduces uncertainty and risk; IT therefore prefers to minimize change.

IT traditionally minimizes risk and manages uncertainty through formal processes. Methodologies such as ITIL give IT a handle on change; unfortunately, it does so by introducing bureaucracy and friction. In addition to its reputation for outright failure, IT has thus gained a reputation for slowness and lack of responsiveness. As a result, industrial IT lacks important characteristics that define a conversational medium. It has neither the ability to move quickly nor to change direction quickly. It lacks a mechanism for incorporating feedback from the rest of the company and, more importantly, for helping the company do the same with its customers.

Some IT organizations, though, have begun adopting an interrelated set of new practices. Together, these practices have the potential to compose the digital medium post-industrial businesses need in order to self-steer through empathic conversations. Chief among them are:

- Agile
- DevOps
- Cloud computing
- Design thinking

These practices all share a cybernetic model of control. This chapter examines them in turn, exploring the specific ways each one helps digital businesses improve service through feedback-based adaptation. It then describes how they come together to create a unified, empathic, conversational medium.

Agile

Agile means "quick and light in movement." Agile arose out of frustration with traditional, heavyweight, *waterfall* software development methodologies. The software industry had become infamous for projects that ran over budget and schedule while failing to deliver what customers really needed. Software project management generally followed an approach one might describe as anti-cybernetic. It was considered important to "get the requirements right up front." Development was delayed until you understood and expressed the project's requirements in full detail. Testing was delayed until development was believed to be complete.

The waterfall model suffered from key conceptual shortcomings. Requirements are very difficult to get right without feedback. First, it's difficult for users to understand and articulate their own needs. Second, it's hard for designers to accurately understand what users are trying to say. The design industry's use of prototyping reflects their recognition of these realities. Finally, even if a user understands what they need and can communicate it to a designer, that need is likely to change to some degree by the time it's been fully implemented.

The History of the Agile Movement

In February 2001, 17 early adopters of so-called "light" software development methodologies gathered at the Snowbird Ski Resort in Utah. They came together to seek common ground between their various approaches. The result is generally considered to be the formal birth of the Agile software movement. It took the form of a manifesto that expressed four main values:

- Individuals and interactions over processes and tools
- Working software over comprehensive documentation
- Customer collaboration over contract negotiation
- Responding to change over following a plan

Since then, the Agile movement has manifested through a variety of specific practices. Some, such as Extreme Programming and Scrum, were represented at Snowbird and have continued through to today.

> Others have evolved in the intervening years. They all share key characteristics that make Agile the grandfather of cybernetic approaches to delivering software.

Waterfall projects are almost inevitably doomed to build things that don't properly match users' needs. Even in the case where an up-front design succeeds, the sequential approach to development and testing generates waste, adds complexity, and compounds errors. Testing can't actually happen "after" development. Bugs must be fixed and then retested. Fixing bugs is part of development. Testing is by definition an iterative and circular process.

Not starting testing until late in the process bunches up large numbers of bugs. Those bugs apply to months' worth of code changes. Debugging and fixing them is complex and time-consuming. Some of this complexity comes from trying to analyze code written a long time ago, while some comes from trying to determine which of many code changes introduced the bug.

Agile methodologies combat the shortcomings of waterfall by breaking development into shorter iterations. Each iteration contains a design/development/testing cycle. It ends with presentation of working software to users. The benefits of this approach all have to do with speeding up the feedback loop. Testing happens more quickly after code has been written. Bug-fixing addresses smaller, more manageable change sets. Users see the results of development sooner and more often. They have the chance to say "that's not quite what I meant" with less painful impact. Development teams are thus less shy about getting feedback and making changes. The result is a process that lets a software organization steer its way to genuine customer value.

CONTINUOUS INTEGRATION

Some Agile teams go even further in integrating testing deeply into the development process. Unit testing makes test-writing part of the developer's job. With unit tests, tests get run every time the software gets built rather than either at the end of the development cycle or the end of an iteration. Continuous Integration (CI) drives this approach to its ultimate conclusion by ensuring the software gets built and tested every time any developer commits a change. CI shrinks the distance between change and feedback to the smallest possible increment. It makes identifying, diagnosing, and repairing bugs as simple as possible by drastically constraining the amount of change being tested at one time.

Agile understands the importance of testing as a feedback mechanism. Without continuous, specific information about the gap between expected and actual in the past, steering into the future can't happen. Scrum uses the practice of retrospectives to take team-level self-steering even further. A retrospective is a regularly scheduled pause in the iterative development cadence to gather feedback about how well the team's process is working. Retrospectives give every team member the chance to reflect on and discuss what went well, what didn't go well, and what could or should be adjusted in future iterations.

SELF-ORGANIZATION

In order to maximize user feedback, Agile emphasizes integrated teams that foster communication and collaboration across disciplines. Testers and product owners, both of whom represent customers' needs and priorities, participate as team members rather than third parties. Some technical teams even integrate design and marketing representatives. Incorporating diverse perspectives lets Agile teams understand and resolve gaps between desired and actual results more quickly, which enables them to accomplish better work more efficiently.

Agile also emphasizes the power of self-organizing teams. Breaking from the Taylorist industrial tradition, self-organizing teams rely on their members to cooperatively define as well as execute work. Teamwork happens through conversation rather than via an externally managed assembly line. Just as with any cybernetic process, intra-team feedback improves Agile teams' ability to cooperate in the face of change. It lets them self-steer by reconfiguring themselves as needed without losing coherency or effectiveness.

BEING VERSUS DOING AGILE

The core of Agile is agility. Done right, it lets teams leverage the power of cybernetics to continuously deliver value in an environment of uncertainty and change. Techniques such as sprint demos and CI maximize feedback. Iterations and backlogs maximize the ability to change direction in response to feedback. Cross-functional collaboration maximizes the ability to listen accurately.

At their best, Agile teams function as foundational elements of self-steering organizations. Their cybernetic practices let them flex with minimal friction. Their cross-functional structure lets them converse with the rest of the corporation. It also helps them have the internal empathic conversations they need to participate in the corporation's unity of purpose.

Unfortunately, IT organizations used to industrial thinking often unconsciously fall prey to the temptation to coopt Agile's post-industrial operating model. Daily standups, instead of being a technique for quick feedback and problem solving, become a rote reporting tool. Burndown charts become a micromanagement hammer, shifting the focus away from delivering useful value and toward finishing assigned tasks on time. Sprint demos become an opportunity to tell users what the team has done rather than a chance to listen to feedback in order to validate the accuracy and continued usefulness of the team's work.

Retrospectives have the potential to be a wonderful, second-order cybernetic mechanism. When used properly, they can help teams avoid getting locked into habitual methods. They keep everyone's focus on the question of whether the team is accomplishing its underlying goal: continuously responding to changing customer needs.

Unfortunately, more often than not, retrospectives fall victim to the "not enough muffins" syndrome. Instead of soliciting feedback about the efficiency with which the team is pursuing its goals and exploring ways to self-correct, participants content themselves with complaining about random obstacles. For a team with a tradition of bringing in food for team breakfasts every Monday morning, not having enough muffins may be a valid complaint. It doesn't, however, address questions about the team's deeper purpose.

In order for Agile to successfully support self-steering, its practitioners must adopt a post-industrial worldview. Agile's challenges are instructive for all of the would-be components of the digital medium. IT needs to keep its focus on the goal, which is self-steering through conversation. It must continually evaluate and improve its ability to support that goal. It must put the ends before the needs. It must learn to view all techniques as provisional and subject to adaptation through their own cybernetic process. "Are we being Agile?" must always remain primary over "Are we doing Agile?"

DevOps

Agile emphasizes delivering value to customers above all else. By itself, though, it doesn't fully address the problem of getting the potential value generated by development into customers' hands. This gap reflects the preservice era, when software was primarily delivered as a product. A software company would build and ship an application with the expectation that the customer was responsible for operating it. The software was done when it had been designed, developed, tested, and burned onto a CD. Installing the software from that CD onto a server, adding memory or upgrading operating systems in order to be able to run it, and figuring out what had gone wrong when it failed were all left to the customer.

Post-industrialism has impacted software just as dramatically as any other area of the economy. Software-as-a-Service (SaaS) means that the same company that builds an application also operates it on their customers' behalf. Functionality and operability become inseparable in customers' minds. They judge a software service as much on the basis of performance, availability, and security as they do on whether it helps them accomplish their desired tasks.

DevOps seeks to complete the equation started by Agile and to address the challenges SaaS poses. The name is a portmanteau of *development* and *operations*. It reflects an understanding of the need to unify functional and operational concerns.

The industrial approach to IT worked by creating silos between specialties and using rigid mechanisms to govern communication between them. Not only did development and operations live different lives; operations subspecialties such as networking, database management, and security did as well. Delivering functionality and operability required navigating multiple layers of separation. Each layer added time, effort, and misunderstanding. Each suborganization viewed the others with suspicion: "they don't understand the importance of ..."; "they expect us to figure out ..."; and so on.

Deeply siloed IT creates waste by requiring bureaucratic communications processes that don't add value. Forcing development to navigate multilayered approval processes in order to deploy code slows that code's availability to customers. Silos within operations exacerbate the problem. Because each subspecialty is measured based on the same metric—stability—they are incentivized to resist change not just coming from development but also from each other. The result is friction built upon friction.

DevOps recognizes that change-averse IT operations is incompatible with post-industrial business. It seeks to dissolve the dichotomy between quality and speed, and to overcome the impedance mismatch between development and operations. Unlike operations, development is incentivized to generate change, not avoid it. Companies pay developers to write code. When you write code, you are either adding, updating, or removing something. Every line of code you write changes something.

SaaS drives a similar focus on speed into operations. New applications require new infrastructure. User traffic that spikes by the day, hour, or minute requires elastic infrastructure. Customer-visible operational problems require fast resolution. SaaS replaces the need not to break things with the need to make things happen in response to environmental change.

THE THREE WAYS

In their book *The Phoenix Project* (*http://amzn.to/1K6DT3n*), Gene Kim, Kevin Behr, and George Spafford describe DevOps as consisting of *Three Ways*. Tim Hunter concisely expresses the *Three Ways of DevOps* as *Flow, Feedback*, and *Continual Experimentation and Learning* (*http://bit.ly/1K6DUVc*). This triad clearly expresses DevOps as an extension of Agile's cybernetic model to the end-to-end software value stream.

Flow

Flow comprises techniques to remove waste from the process of delivering concrete value to customers. It redefines *done* in the language of service rather than products. Many organizations that call themselves Agile define *done* as sprint tasks completed, unit tests passing, or projects signed off by testing and product owner. DevOps redefines *done* as "available to customers and operating satisfactorily."

Flow includes three key practices:

- Cross-silo collaboration
- Automation
- Continuous Delivery

Cross-silo collaboration uses human communication to improve quality and reduce waste. When developers, system and database administrators, and security engineers all talk to one another, they can see problems and find solutions in the spaces between their domains. The result is improved quality, achieved more efficiently. On a more subtle level, cross-silo collaboration also improves flow by generating empathy and dissolving suspicion. People are naturally more prone to remove roadblocks and help each other when they have mutual trust and understanding.

Automation seeks to reduce manual tasks that unnecessarily consume time and cause errors. There's no good reason to manually configure 500 servers in the exact same manner. There's even less reason to spend hours or days debugging a production problem that was caused by one server having been configured slightly differently from the other 499.

Infrastructure as Code

Automation leverages lessons from Agile development. It uses high-level configuration languages that bring the benefits of encapsulation, abstraction, and reuse to system administration. It treats *infrastructure as code*, turning computing environments into abstract configurations that can be version controlled, unit tested, and continually integrated. As a result, they can be created, changed, and replicated quickly and safely, with guaranteed consistency.

Configuration automation, for example, uses textual descriptions to represent desired server state. Agents running on each server interpret these descriptions to ensure that the proper applications, libraries, and files are present. This technique makes it possible to guarantee that every server of a given type is identically configured. It can speed development and testing, and reduce errors, by ensuring that development, test, and production servers all share the same configurations.

IT normally delivers changes to customers in batches called *releases*. So-called "big-bang" releases delay value delivery by weeks, months, or even years. Just as with any other large-granularity activity, batched releases increase the complexity, difficulty, and risk of deploying changes to production environments. Risk-averse IT organizations respond counterproductively by trying to reduce the number of releases, thus locking themselves into a vicious cycle by making them even more risky.

Continuous Delivery (CD) does for deployment what CI does for integration testing. It uses comprehensive automation and rigorous testing to enable immediate production deployment of any change deemed acceptable, regardless of how small. A release might be as small as a one-line bug fix.

By reducing batch size, release latency, and errors due to manual processes, and by guaranteeing comprehensive testing, CD reduces risk and increases confidence. It leads to fearless releases. With the confidence to release any change, any time, the organization gains the flow needed to let it respond immediately to its customers. That flow lets companies have the most intimate conversations at the most valuable times.

Many companies with seasonal business cycles implement production freezes: times during the year when only emergency changes can be deployed to business-critical systems. The rationale for this approach is understandable: if you're an online retailer, the day after Thanksgiving is the day you least want to break your website. Unfortunately, it also minimizes your ability to respond to your customers when it's most important. Fixing a bug or releasing a desired new feature on January 10 instead of November 28 doesn't benefit anyone.

When they first hear about CD, marketing and business operations often respond with alarm. They envision their customers being inundated with uncontrolled, unchaperoned change. Decoupling technical deployment from customer visibility is a key component of CD. Through mechanisms such as feature flags and segmented releases, CD frees the business to truly control its conversation with customers.

Feature Flags

Feature flags make it possible to precisely control which features are visible and when. If marketing wants to reveal a new feature at 12:01 AM on Christmas morning, they don't need IT to deploy that code at 12:01 AM. They can simply flip a virtual switch. Segmented releases turn what used to be known as *final* or *golden* into controlled experimentation. Marketing can expose a feature to specific demographics, or use A/B testing to release multiple versions of a feature simultaneously.

Feedback

Flow lets IT organizations efficiently deliver functionality and operability to customers. Cybernetic conversations require equally efficient feedback. In the context of DevOps, feedback seeks to provide information back to development and operations as continuously as possible. Its goal is to generate a frictionless loop that lets IT hear its customers quickly and accurately across silos.

Monitoring provides the visibility necessary to process feedback. IT should treat it as an integral part of application and system design, implementation, and operations. It must be possible to ask questions on multiple levels. What is the state of the infrastructure? What is the state of the application? What is the state of the users' behavior?

From a DevOps perspective, teams across silos need the ability to listen to each other's signals. Multilevel monitoring lets them see things from each other's perspectives. For example:

- Users are abandoning our site because it's too slow

- It's too slow because we don't have sufficient infrastructure

- We don't have sufficient infrastructure because adding more is expensive and time consuming

- An on-demand cloud might make infrastructure cheaper, easier, and faster to provision

The delivery lifecycle itself is a cybernetic process. As such, it also requires the ability to process feedback. Information radiators offer global visibility into flow:

- What is the state of the current build?

- Why did a given build fail?

- Where in the pipeline is a given feature?

- Which customer segments have access to a given feature?

Continuous Experimentation and Learning

The combination of flow and feedback creates an efficient cybernetic process that lets IT shift its focus from maintaining the status quo out of fear to fearlessly enabling continuous change and experimentation. DevOps lets IT unfreeze digital systems, transforming them into a continually flowing fluid. It lets IT unleash software as a dynamic business tool that digitally infused companies can use to deliver service and respond to disruption.

DevOps frees post-industrial businesses to fully integrate software into their self-steering conversations. Marketing and business operations can use CD to turn features on and off with minimal latency, and multilevel monitoring to understand the relationship between changes and their customers' responses. They can thus treat software not as something to which they must commit but rather as something they can continually reshape through experimentation and learning.

Cloud Computing

Part of IT's challenge stems from the fact that provisioning computing resources is expensive and slow. As with any physical object, servers, storage arrays, and network devices have long lead times. They must be specified, ordered, built, shipped, and installed. Deploying a new application, or scaling infrastructure to support increased usage, requires high-latency physical intervention. The high cost of modifying infrastructure contributes to IT's resistance to change.

Cloud computing makes it possible to treat computing resources as an on-demand utility. It works by virtualizing physical resources, and providing web and API-based provisioning interfaces. By abstracting the underlying physical reality, cloud lets users dynamically provision and deprovision resources for themselves. They can even integrate cloud provisioning APIs into their automation pipelines to create an elastic computing substrate that responds to change in real time.

Cloud lets organizations consume IT based on need, pay for it based on consumption, and delegate its management to the provider. In the process, it transforms IT from a relatively static capital expenditure to a highly dynamic operational expenditure. In both technical and financial terms, cloud more closely aligns IT with the ebb and flow of the business.

Cloud computing addresses multiple layers of IT. *Infrastructure-as-a-Service (IaaS)* turns physical infrastructure into a utility. *Platform-as-a-Service (PaaS)* turns application execution environments into a utility. *SaaS* turns applications themselves into a utility.

All three layers share the "as-a-Service" moniker. This common name reflects the impact of service, infusion, and disruption on IT itself. While cloud makes IT more Agile by increasing provisioning speed and reducing sunk costs and management complexity, it also has a more subtle and profound effect. When IT resources are accessible through the same digital interfaces as any other application, it becomes feasible for nonoperations groups to manage their own resources. IaaS and PaaS let development teams provision their own application substrates. SaaS lets nontechnical organizations provision their own applications.

Cloud's ultimate effect is to remove IT as a bottleneck to self-steering. On one level, it lets IT flex in response to feedback at the same rate as the organizations using those resources. On a deeper level, it begins to dissolve the separation between IT and its users. As part of self-steering, organizations need access to the digital medium in order to converse with one another and with customers.

The more flexible and direct that access is, the more efficient those conversations can be.

MICROSERVICES

By providing low-friction access to the digital medium, cloud computing allows systems and applications to proliferate. *Microservices* are the ultimate expression of this effect. Microservices decompose large applications into small, loosely coupled grains of functionality.

By holding units of functionality at arms length from each other, microservices enable more continuous, lower-risk change. Agile and Continuous Delivery use smaller batch sizes to increase speed and quality simultaneously. Microservices provide a similar effect at the level of application architecture and organizational structure.

Microservices work by reducing the scope of concern. Developers have to worry about fewer lines of code, fewer features, and fewer interfaces. They can deliver functional value more quickly and often, with less fear of breaking things, and rely on higher-order emergent processes to incorporate their work into a coherent global system.

In order for microservices to work, though, operations needs a similar conceptual framework. Trying to manage an entire universe of microservices from the outside increases the scope of concern instead of reducing it. The solution is to take the word *service* seriously. Each microservice is just that: a software service. The team that builds *and* operates it need only worry about it and its immediate dependencies. Dependent services are customers; services upon which a given microservice depends are vendors.

Microservices thus represent a new IT organizational model as much as a new architectural model. Seen this way, they leverage the power of Conway's law (*http://bit.ly/1K6ELVO*), named after Melvin Conway. Conway was a computer programmer who observed that "organizations which design systems ... are constrained to produce designs which are copies of the communication structures of these organizations."

Conway's law tells us that software architectures and the organizations that make them mirror each other. IT can leverage this effect to everyone's benefit. Microservices map well to so-called *two pizza-sized*, interdisciplinary teams. These teams take responsibility for the entire service delivery lifecycle for their particular microservice.

Microservice-oriented organizations shift IT architectures, processes, and interrelationships from a complicated-systems model to a complex-systems model. Making that transition lets IT better respond to post-industrial business challenges. It lets different parts of the organization, as represented by different microservices, fluidly adapt to one another and to external change. By taking the *service* in microservice seriously, it also weaves an empathic, conversational mentality deeply into the fabric of IT.

Digital infusion makes the relationship between IT and the business it supports all the more important. One could extend Conway's law to state that digital businesses are constrained to deliver service in ways which reflect the structure and activity of their IT organizations. Using microservices to give itself a more organic structure increases IT's ability to self-steer through flexible, scalable, internal conversations. That ability in turn drives improved self-steering capabilities for the organization as a whole.

Design Thinking

Digital infusion is having a profound effect on the role of design in IT and software services. As the digital realm becomes ever more central to our lives, the design of its interfaces becomes ever more important to people's quality of life. Design brings its own cybernetic sensibility to delivering products and services. The design community has encoded this sensibility into a set of practices and principles known as *design thinking*.

Design thinking is built upon four foundational practices:

- Empathy
- Ethnography
- Abductive thinking
- Iterative user testing

Empathy makes the customer's perspective on a problem the starting point for all design activity. It reflects the philosophy of user-centeredness. No matter how beautiful a design solution, it doesn't actually solve anything if it doesn't work from the user's point of view. An elegant chair that no one can sit in is a user-centered design failure.

Ethnography is a disciplined process of nonjudgmentally observing users within their own realms. Without ethnography, designers risk unconsciously imposing their own biases instead of truly seeing the problem from the customer's perspective. Many design teams have described their experience of starting a project believing they were trying to solve a certain problem. After conducting ethnographic research, they realized the real problem was completely different. Having engaged in that research saved them from wasting everyone's time building the wrong solution.

Abductive thinking is the process of finding creative solutions where there are no *correct* or *best* ones. Abductive thinking succeeds in situations where analytical engineering fails. It strives for designs that are practical as well as beautiful and inspiring.

Iterative user testing forces designers to repeatedly test and revise their beliefs about a solution. Design thinking views the development of a solution to a problem as the starting point, not the conclusion. User testing exposes proposed designs to the harsh reality of usage in the form of prototypes. It treats users' experience of those prototypes as feedback.

Repeated revision and retesting leads to successively better designs. Ethnography, user testing, and iterative solution discovery incorporate feedback into the essential process of design. As design thinking evangelist Elina Zheleva (*http:// bit.ly/1K6EUIO*) puts it, design follows a circular *understand-act*.

SERVICE DESIGN

Design thinking expresses designers' sensibilities in a form that can be applied to problems beyond traditional disciplines such as graphic and product design. In particular, design thinking has tremendous insights to offer to the creation of services. By engaging in empathy, ethnography, abductive thinking, and iterative user testing, designers can create services that genuinely help customers accomplish their goals.

Service design applies the principles of design thinking to the design of services. It centers its practice around the customer journey, which represents customer's unfolding experiences over time across all of a service's touchpoints. Applying design thinking to the customer journey helps create experiences that are satisfying and coherent rather than challenging and disjointed.

Service design also addresses service fulfillment. It uses a technique called service blueprinting to chart the relationship between so-called *front-stage* and *back-stage* activities. However well-designed a kiosk interface is, it can't provide a satisfying experience if it doesn't properly integrate needed back-office information. All of the component human and computer processes needed to generate that information must mesh with one another.

Customer journey maps and service blueprints help service organizations understand themselves and their customer interactions across silos and layers, across physical and virtual interfaces, and across human and computerized processes. Digital infusion necessitates seamless integration across all of these dimensions.

A self-steering organization needs to think in terms of relationships and systems. It needs a way to visualize its internal conversations and how they enable or distort its customer conversations. Through customer journey maps and service blueprints, service design has the potential to provide such a mirror.

Unifying Design and Operations

Post-industrial business involves operating a continuous "listening loop" through which companies can respond efficiently and accurately to customer needs. In order to maintain their viability through self-steering, organizations need to map that external listening loop to a similar set of internal conversations. Digital infusion means that internal and external conversations all happen through software-enabled service. Twenty-first-century business thus relies on IT to enable the continuous design and operation of responsive digital services. These services provide the medium through which cybernetic conversations flow, both between companies and their customers, and among employees and groups within a company.

The purpose of the digital conversational medium is not merely to deliver software, but rather to enable continuous, adaptive value co-creation. Whether it involves a company operating a website on behalf of its customers or the finance department operating a microservice on behalf of the project management department, *service* is the key word in all cases. In the post-industrial era, service unfolds through the unification of design and operations.

In order to fulfill its purpose, a software service must work on multiple levels. It must provide suitable functionality. That functionality must be usable, whether through an interactive interface or through an API. It must be operable so that its customers can access it when they need it and rely on it for stability, security, and so forth. It must meets customers' needs throughout their journey. Finally, it must adapt to meet changing needs. If, for example, the project management department needs finer-grained project-cost information, the finance department has to be able to update its service to provide that new information.

These requirements apply to all services, whether internally or externally facing. To fulfill its role as a conversational medium, IT therefore must address all aspects of fitness for purpose. It needs to incorporate the capabilities provided by methods such as design thinking, Agile, DevOps, and cloud computing into a coherent practice. This practice uses a unified set of principles to guide itself in finding comprehensive answers to the full suite of questions that define digital service needs.

The fundamental principles that guide IT as a digital conversational medium include:

- Design for service, not just software

- Minimize latency, maximize feedback

- Use operations as input to design

- Seek empathy

Designing for service means designing both for the whole customer group and the whole organization. Designing for the whole customer group starts by understanding customers' goals. It identifies the entire journey through which they interact with the service in order to accomplish those goals. It also identifies the larger context that surrounds their interaction with a given service. Finally, understanding the whole customer group requires considering their needs beyond the obvious, utilitarian level. Productivity arises not just from efficiency but also from satisfaction and happiness. Making people's lives better through service thus contributes to meeting practical goals.

The co-creative nature of service means that the service organization also must be a design target. Those designing a service need to ask the same questions about internal, operational users as they do about customers. Most importantly, they must address the interrelationships between internal and external goals and journeys.

Conversational quality depends on efficient information exchange. Communicating by letter across continents, for example, is harder and slower than speaking in person. The digital conversational medium needs to minimize latency in the exchange between service providers and customers. Together, Agile, DevOps, and cloud computing serve this purpose. Agile minimizes the latency between discovering a need and implementing it. DevOps minimizes latency in the end-to-end delivery, understanding, and discovery process. Cloud computing minimizes computing-resource deployment latency.

Continuously delivering functionality is only half of the conversational process. Talking without listening doesn't contribute to responsive service. When done properly with a truly cybernetic sensibility, design thinking, Agile, and

DevOps complete the conversational equation. They continuously expose the service organization to accurate, thorough feedback, and provide mechanisms such as user testing, sprint demos, and information radiators that maximize the organization's ability to internalize and process that feedback.

LeanUX

LeanUX is a wonderfully cybernetic practice that integrates Lean Startup, design thinking, and Agile. While design has always taken an iterative approach, it traditionally iterated its way toward a static solution. Eventually, one needed to produce a finished chair, alarm clock, poster, or building. User interface design generally takes the same approach. It follows a design process with the intention of creating a heavyweight set of deliverables that describe the finished UI. Changing the UI involves undergoing a similarly heavyweight process all over again.

Digital businesses can't afford such a product-centric process. Agile and Continuous Delivery give the ability to constantly iterate code in response to changing customer needs. They need to be able to take the same approach to UI design.

LeanUX attempts to heal the split between industrial design methodologies and post-industrial software methodologies. LeanUX values working user interfaces over traditional design deliverables. It emphasizes experimentation, iteration, and learning over freezing functionality. It leverages the insights of Lean Startup to maximize learning through feedback while minimizing the investment and latency required to acquire feedback.

LeanUX offers an example that points the way toward a unified digital medium. It does for design and software development what DevOps does for development and operations. There are potentially fruitful opportunities to explore the intersection between LeanUX and DevOps, especially in the context of service design.

Businesses normally treat operations as an output of design. The job of IT operations is to run the code produced by design and development. In order to empathize, though, one must be able to hear. In order to hear, one needs information from operations. Operations thus becomes an input to design.

Operational feedback comes from multiple sources on multiple levels, including:

- Infrastructure and application monitoring
- User behavior monitoring
- A/B, canary, and demographically targeted testing
- Analytics
- Customer support
- Social media

An effective conversational medium incorporates them all. A user interface change could annoy users either because it degrades performance by increasing server load or because it makes the application harder to use. Problems can become apparent through monitoring dashboards or through users complaining on Twitter. Correlating feedback across business and technical layers is key to accurately diagnosing and fixing problems, whether they be caused by infrastructure problems or undiscovered customer needs.

Empathy is a cybernetic process of understanding through conversation. It needs to be both the foundation and the goal of the conversational medium. On the one hand, empathy should inform the pursuit of each of the other foundational principles. On the other hand, those principles should all be approached as opportunities to develop further empathy.

CONTINUOUS DESIGN

IT's new purpose is to help companies and their component parts design and operate software services. No longer can it content itself with running systems and responding to technical requests. Where IT used to be in the business of running things and maintaining stability, it now must enter the business of enabling change. Ultimately, post-industrial IT's new mandate consists of delivering the capability for continuous design.

Design typically concerns itself with what comes next. It focuses on conceiving new solutions to current problems. Operations, on the other hand, concerns itself with what's happening now. Its purpose is to run and maintain whatever solution was created to a previously understood problem.

The cybernetic model of control breaks down the divisions between these two modes of work. It unifies them through feedback. Feedback continually exposes gaps between the actual and the desired. In the process, it creates never-ending opportunities to co-create value by helping people solve problems.

Empathy drives digital businesses to use conversation as a basis for action. It gives them the ability to hear the feedback that operations provides and exhorts them to respond to what they hear through redesign. Far from being something soft or weak, empathy drives economic sustainability by creating the impetus to design—or in other words, responsively operate—truly useful service capabilities.

The industrial product model uses design to generate solutions. Marketing then convinces customers of the usefulness of those solutions. Operations produces artifacts to meet the demand that marketing has generated. A post-industrial approach to design, by contrast, uses it to generate conversations rather than complete them.

As design consultant and researcher Thomas Wendt explains in his book *Designing for Dasein*, the meaning of a digital service depends not just on the intentions of its creators but also on how people use it. Facebook was designed as a way for college students to connect with their friends. It has evolved into, among other things, a platform for catalyzing political action. At the same time, designed solutions change the very problems they were created to solve. Facebook has deeply influenced the way people see themselves as individuals and as social beings. It has transcended its role as a medium for sharing experiences to become a medium for having them.

Complexity further compromises the ability to create finished solutions. We can never fully know how our designs will work until real customers use them in real operating environments. Facebook has to grapple with social and political issues (*http://bit.ly/1G8XM7P*) that Mark Zuckerberg could never have conceived of while writing the initial version in his college dorm room.

Post-industrialism transforms design into a circular process of continual learning and repair. Repeatedly responding to the next gap between actual and expected becomes the essence of what it means to be in business. Operations becomes design, and design becomes operations.

Design as a continuous, circular process is actually inherent in its very definition. Nobel Laureate and pioneering cognitive scientist Herbert Simon, in his seminal book *The Sciences of the Artificial* (*http://amzn.to/1EHezt3*), defined design as "courses of action aimed at changing existing situations into preferred ones." This definition has several important implications:

- Design is not restricted to visual disciplines

- Design fundamentally involves change

- Design operates in the gap between actual ("current") and expected ("preferred")

- Design does not strive for right or even good solutions, but only relatively better ones from the user's perspective

- Design's lack of objective finality maintains the openness that allows and encourages further change (what was preferred is now current)

SELF-STEERING AS CONTINUOUS DESIGN

Digital businesses design and operate service capabilities to help customers accomplish their goals. In order to co-create value, they must align their internal structure and activity with that of their customers. Value co-creation arises from structural coupling between the service provider and its environment. If customers need help using an application, for example, the service organization must structure and manage itself to be able to provide help when and how it's needed.

This structural coupling must, of course, continually adapt to environmental change. Change comes from ever-evolving customer needs and market realities. Companies also contribute to environmental change through their own design process. Customers, markets, and companies continuously perturb one another.

In order to maintain their viability by adapting to environmental perturbation through self-steering, digital businesses must design and operate themselves just as they design and operate the services through which they interact with their customers. Continuous design is an autopoietic process. As part of designing/operating a service, a digital business must design/operate itself. To understand how to design/operate itself, a business must understand the design/operations of the service through which it maintains its viability.

In order to deliver a digitally infused taxi service, for example, Uber must do more than just design and operate a mobile app. It also must design and operate itself as an organization with capabilities such as calculating driver tips and conducting background checks. In the process of operating its service, Uber learns how well the service *and* its internal systems work.

Feedback from internal and external operations leads to further internal and external design. Drivers may complain about being under-tipped. Customers may complain about not feeling safe due to inadequate background checks or about having to wait because of inaccurate arrival estimates. In order to address these complaints, Uber might need to redesign its service, mobile applications, and internal operational procedures.

Cybernetics challenges the solidity of the boundaries between inner and outer, production and consumption, and acting and responding. Autopoiesis defines business viability as continuous organizational adaptation to markets and customers. Continuous design reflects the cybernetic model on two levels. First, it treats design and operations as an inseparable, mutually influencing pair. Second, it treats the design of internal and external service relationships as a similar process of mutual influence and reflection.

From Design Thinking to DevOps and Back Again

Simon's definition of design as "changing existing situations into preferred ones" reveals it as something that goes beyond traditional visual disciplines such as graphic, industrial, or web design. His definition gives us a lever we can use to reimagine IT itself as a form of design. Cybernetic methods such as Agile, DevOps, and cloud computing free IT from the need to map dynamic business needs to static technical systems and processes. As a result, IT can transform its view of itself from a source of friction into an agent of responsive change.

The capability to deliver continuous, empathic change is the defining characteristic of a digital conversational medium. The word *medium* means "an intervening substance, as air, through which a force acts or an effect is produced." Post-industrial IT's purpose is to make digital conversations between companies and their customers natural and effortless, the way water makes it possible for fish to swim. As an autopoietic medium, fulfilling its purpose also means simultaneously enabling equally effortless internal conversations.

IT as conversational medium transcends specific techniques. It represents more than just designing and operating services for customers or employees. It is not defined by the specific tools and practices that make it up. Its deepest value comes from infusing entire organizations with design thinking.

When the capacity for responsive change becomes frictionless and universally available, everyone can approach their work as continuous service design. Service characterizes all the relationships that define digital business. Post-industrial companies co-create value with customers through service. That co-creation relies on mutual internal service. Designing and operating service for others' benefit becomes the common driver for all activity at all levels of the organization.

IT's ultimate goal is, like water for the fish, to disappear. In a highly effective digital business, employees focus on continuously transforming empathic listening into acting. They conduct their daily work in order to change existing situations into preferred ones. They take it for granted that identifying the gap between current and preferred, and closing that gap, both happen via digital means. They no longer need to step out of the continuous design mindset in order to translate between service design thinking as a process and IT as the means for accomplishing that process.

4

Designing for Failure, Operating to Learn

Digital infusion confronts post-industrial businesses with ever more complex environments. Traditional boundaries, such as those between back-office and front-office functions,[1] break down. Even the idea that IT is separate from the rest of the business, and from the means of customer engagement, comes under scrutiny. Brand management becomes exponentially harder when it has to reflect the operation of an entire service organization.

The continuously self-designing organization represents the triumph of complexity over complicatedness. Its autopoietic process is the ultimate expression of emergent structure. The organization and its components continually create, adapt, and rearrange one another. The shape of its structure becomes a dynamic process that unfolds over time. The history of that process reflects the history of the organization's interactions with its customers and the market.

The flexibility that arises from continuous design dooms attempts at complicated-systems control. The digital conversational medium allows organizations to conduct intimate conversations with their customers. In the process, however, it forces digital businesses to reimagine their understanding of success and failure.

Industrial-era control mechanisms strive to identify and eradicate opportunities for failure. The ideal production workflow is error free. It turns out products that are free from defects and identical to one another. In addition to banishing error, this approach also banishes innovation and adaptation. Particularly in com-

---

1 "Front-office" and "back-office" correspond to what IT typically refers to as Systems of Record and Systems of Engagement.

plex environments where component-level failure is unavoidable, it results in brittleness instead of robustness.

Cybernetic control mechanisms, on the other hand, incorporate failure within themselves. They make the calculation and correction of error part of steady-state operations. They provide a basis for rethinking the meaning of failure in a way that can accommodate the sloppiness of complex systems. This capability for accommodation is critical to leveraging the resilience that accompanies that sloppiness.

The cybernetic model of control that underlies the digital conversational medium lets us trade the precise robustness of complicated systems for the sloppy resilience of complex systems. In the process, it defines the relationship between success and failure in less dualistic terms. From a second-order cybernetic perspective, one might rephrase "calculation and correction of error" as "recognition of and response to difference." The ability to adopt a more unified understanding of success and failure is indispensable for successful post-industrial management.

Redefining Success

Complexity confounds our traditional understanding of the meaning of success and failure. Instead of a binary decision at a point in time, it becomes a dynamic, evolving process that can be evaluated only in hindsight. We generally equate *success* with *correctness*. Post-industrialism brings that equivalency into question.

These days failure is all the rage. Blog posts and tweets announce the good news: failure is valuable, failure is necessary, failure should be encouraged, Google developed all of its best services out of failures, and so on. The idea that failure is good, though, seems to imply that somehow it leads to success. If that's the case, is it really failure anymore? What does it really mean to say that "failure is good"?

By itself, failure is anything but good. Making the same mistake over and over again doesn't help anyone. Failure only leads to success when we learn from it by changing our behavior in response to it. Even then, it's impossible to guarantee the accuracy of any given response. Its validity can only be evaluated in hindsight. The environment to which you're trying to adapt doesn't stop changing just because you've declared victory. Yesterday's success can turn into tomorrow's failure.

We need a new definition of failure that shifts our focus from momentary events to unfolding processes. This shift is especially important in the context of complex systems that evade traditional control. Component-level failure is inevitable in complex systems, yet the systems themselves can still thrive and even improve. Conversely, component-level events can combine to cause systemic breakdowns without themselves being considered failures. Once again, failure is in the eyes of the future beholder.

We seem to be stuck in a catch-22. On one hand, we can't be sure our actions won't make things worse. On the other hand, inaction isn't an option; the situation arose in the first place because our current state is unsatisfactory. To resolve this conundrum, we need a less dualistic, binary approach to success and failure. Complexity forces us to redefine success as "a useful conversation with one's environment."

Post-industrial IT provides the conversational medium by which self-steering organizations navigate uncertain, unknowable, continuously changing environments. Put another way, it helps them have useful conversations with those environments. Because complex systems are resistant to industrial management techniques, twenty-first-century businesses need a new strategy. Instead of trying to tame their environments by engineering out the possibility for breakdowns, they need to develop the ability to continuously repair them.

Success as Conversation

A conversation is "an interchange of information." A useful conversation exchanges information in a way that continually makes sense and offers value to all involved parties. Imagine the following counterproductive exchange:

Speaker 1: Let's have Indian food for dinner.

Speaker 2: I don't like Indian food.

Speaker 1: Do you prefer the Indian place on Grand or the one on Summit?

Speaker 1 isn't really listening to his counterpart. He's not leading the conversation in a direction that has anything to offer to Speaker 2. In fact, Speaker 1 is wandering off into the weeds by way of non sequitur.

By this definition, success is less about what you do at any given point in time than how you process the environment's response to it. The following conversation is perfectly constructive:

Speaker 1: Let's have Indian food for dinner.

Speaker 2: I don't like Indian food.

Speaker 1: Do you like Mexican food?

Speaker 2: I love it!

Speaker 1: I know a great Mexican place on Grand.

Speaker 2: That sounds good. But wait, won't Grand be congested tonight because of the game?

Speaker 1: Good point. There's another good place on Summit.

Speaker 2: Let's go there.

By redefining success in this way, we've enabled ourselves to evaluate our progress in real time. As long as we're (a) speaking, that is, acting by trying something new, (b) listening to the environment's response to our action, and (c) guiding our future action by the response to our past action, then we are succeeding. When we stop engaging in any of those three steps, we have failed.

CONVERSATION AS CONTINUAL REPAIR

Whether it be a vendor and its customer, a finance department and a project management department, or an authorization microservice and a login microservice, conversation always takes place between agents that differ from one another. Regardless of the level of empathy, perfect mutual understanding is never feasible. As one can see even from the simplistic example of the conversation about where to have dinner, breakdowns in understanding are inevitable and continual. Success through conversation is a matter of continual repair.

By its very cybernetic nature, conversational IT enables useful conversation through continual repair. Feedback loops continually correct conversational misunderstanding. In fact, one could think of self-steering as another word for a "useful conversation with one's environment." If self-steering doesn't generate a meaningful conversation, it doesn't help the system stay alive.

The Agile practice of sprint demos perfectly illustrates this mechanism. The sprint demo's ostensible purpose is to *demonstrate* the development team's progress to customers or their proxy. Its real purpose, however, is to get feedback about errors in understanding. It gives customers an opportunity to say "That's not really how we wanted it to work," or "That's not really what we need."

Success and Failure in Complex Systems

Complex systems are enigmatic and rife with failure. Their properties of emergence, resilience in the face of component-level failure, cascading failures, and sensitivity to history make them continually surprise us. Their resistance to top-down management makes conversational *control* the only feasible method. Continuous repair requires the willingness to listen and to be led; in Glanville's words, "to control by being controlled by the controlled."

The futility of trying to control complex systems through traditional means is not without its benefits. People tout the value of failure because it contributes to learning. Instead of trying to keep complex systems from failing, we can use their failures as learning opportunities. The willingness to listen to and be led by failure means that we can build and operate systems and organizations that improve over time, rather than just not degrading.

To use continuous design as a strategy for steering through complexity, we need to augment it with an additional principle: *design for failure; operate to learn.*

Cutting-edge organizations are espousing this principle through a variety of practices, including:

- MTTR over MTBF
- Design-for-fail
- Game days
- Chaos monkeys
- Blameless postmortems
- Operational transparency

These practices typically appear inside of IT. They don't, however, only apply to problems of technical resilience. The challenges of complexity confront digitally infused service as a whole. The entire service organization needs to design for failure and operate to learn. After describing specific practices for steering through complexity in the context of IT, this chapter will examine them from an overall service business perspective.

MTTR over MTBF

Traditionally, IT exerts tremendous energy trying to maximize mean time between failures (MTBF). It strives to make underlying systems (hardware, network, databases, etc.) as robust as possible. This strategy allows higher-level systems and applications to take a naïve approach to failure. They only need to concern themselves with it at their own level. They can assume lower-level failures either won't happen or will be contained and hidden.

Maximizing MTBF doesn't work in complex environments with large numbers of independent, rapidly changing agents. At some point, failure becomes inevitable by simple fact of arithmetic. When you have thousands of systems and are continuously deploying many small changes, trying to maximize MTBF no longer suffices. As Adrian Cockcroft, former Chief Cloud Architect at Netflix, says, "speed at scale breaks everything."

One might think that Netflix's scalability challenges don't apply to normal enterprises. Netflix does, after all, account for 33% of all Internet traffic on a typical Friday night. A large enterprise, though, operates hundreds if not thousands of applications. Those applications run on thousands of production server instances and process millions of transactions against many terabytes of data. By decomposing applications into more fine-grained modules, service-oriented architectures multiply the number of software objects needing management. Netflix's lessons are thus more relevant to most enterprises than one might realize.

The alternative to struggling to maximize MTBF is to shift your strategy toward minimizing mean time to repair (MTTR). If it's sufficiently quick, easy, and safe to repair faults, you become less afraid of them. Continuous Delivery, fearless releases, infrastructure as code, and microservices architectures are all components of an MTTR-centric fault-management strategy.

Continuous Delivery minimizes the latency of change and thus repair. Imagine that a bug has been discovered in production. Without Continuous Delivery, a bug fix has to be attached to a release. The latency with which that release moves from development to production is proportional to the complexity of all the changes contained within it. With Continuous Delivery, on the other hand, the bug fix can be released to production as quickly as a single change can be coded and tested.

Fearless releases are an extension of Continuous Delivery. Having practiced, automated, and wrung the difficulty out of the change deployment process, you no longer fear it. The prospect of having to change production in response to a failure no longer makes you hesitate. On both physical and psychological levels, fearless releases help reduce deployment latency and thus reduce time-to-repair.

Infrastructure as code makes it possible to apply Continuous Delivery and fearless releases to infrastructure as well as application changes. Software-as-a-Service, with its inseparable requirements for functionality and operability, necessitates treating infrastructure and application changes inseparably in terms of quality and speed. Failures happen at all levels of the IT stack; we need the ability to minimize MTTR throughout.

Microservices increase the surface area of code by decomposing applications into small, loosely coupled services. Reducing size and coupling shrinks internal complexity and the scope of impact of change. Combined with Continuous Delivery, microservices can dramatically reduce the duration and cost of failure. Imagine a monolithic application with a monthly release cycle. A bug in production will impact all users for, on average, 15 days. If we assume 1,000 users, the average cost of the bug is 15,000 user-days.

Now imagine a microservices architecture with a Continuous Delivery release process. One service has a bug. That service impacts 1/10th of the user base. Releasing a change to production takes on average one hour. The average cost of this bug is 100 user-hours, or approximately four user days. The combination of microservices and Continuous Delivery has thus reduced the cost of failure by a factor of nearly 4,000!

Design-for-Fail

Complexity makes it infeasible to assume robustness on the part of the systems upon which one relies. Instead, one needs to design systems and applications at any given layer in the IT stack based to assume the possibility of failure at lower levels. Design-for-fail necessitates thinking about fault tolerance at all layers. No longer can application developers confine themselves to thinking about functionality. They must also consider the question of how to provide that functionality in the face of database, network, or other outages.

Design-for-fail impacts not just code but also design. Amazon's use of a microservices architecture for its website means that the site as a whole nearly never goes down. Individual services, on the other hand, experience outages and degraded performance all the time. Amazon has built its user interface to gracefully degrade in the face of service failures. If a given service is misbehaving, the website will remove that service from the user interface. It essentially takes degraded functionality out of production from the user's point of view.

It's this level of design-for-fail that makes the microservices cost-of-failure math truly work. In order to say that a bug only impacts a subset of your customers, you have to be able to render the bug invisible to the rest of your customers. If, for example, Amazon's customer review service is misbehaving and the company removes it from the visible interface, it only impacts customers who want to read reviews during the period of the service outage.

Game Days

Design-for-fail promises the ability to gracefully survive component failure. No design for a complex system, however, can completely prevent visible errors. No matter how well we design complex socio-technical systems, things inevitably go wrong. We need to know how to respond when components and systems fail in order to fix them. What do we do if a database crashes? Perhaps more interestingly, what do we do if the database failover system or the data backup and restore system fail? Most importantly, how do we even know what's liable to fail?

In 1958, Ross Ashby, one of the pioneers of cybernetics, identified a principle he called the Law of Requisite Variety. Simply put, the law states that a control mechanism must be as rich as the system it's controlling. If an IT system can fail in any of 100 different ways, the IT organization needs the ability to recognize and respond to all 100 kinds of failures. In order to develop an appropriately rich failure management mechanism, the organization needs to identify the possible failure scenarios along with adequate responses to them.

Game days evolved as a means of generating requisite variety for IT failure management. Game days simulate real, game-time situations. These simulations intentionally generate failures in test environments in order to find out what might go wrong and to test possible responses. Game day facilitators will do things like shut down databases, disconnect networks, and delete data without telling the simulation participants. These exercises assume that control systems and operational systems can fail. By uncovering and accounting for failure modes in simulation mode, Game days can help prevent real failures in production.

GAME DAYS IN ACTION

The 2008 U.S. presidential election illustrated the power of game days. The Obama and Romney campaigns both relied on volunteers going door to door on Election Day to maximize voter turnout in their favor. The effectiveness of this strategy depended on volunteers' ability to accurately target which doors on which streets to knock.

Each campaign developed software that allowed volunteers to access voter information in order to guide them as they walked about their neighborhoods. The Obama campaign ran game days during the weeks leading up to election day. The Romney campaign didn't finish developing its application until just before the election; as a result, there wasn't time to test it in advance. When Election Day came, the Romney campaign's application melted down. The Obama campaign's application, on the other hand, ran without a hitch. It would be overstating the situation to claim that the different won the election for Obama; it did, however, make a nontrivial difference.

Chaos Monkeys

If we know that our systems are rife with failure, what should we do about it? Should we try to eradicate it even though we know the attempt will likely backfire? Should we ignore it and hope for the best? Or should we instead go looking for failure and try to expose it to the light of day. That way we can learn from it and adapt to it, making our systems better as a result.

Netflix pioneered exactly this approach with its so-called chaos monkey. The *chaos monkey* is an automated software system that intentionally generates component failures in production. The chaos monkey in effect runs game days during the real game.

One of the challenging characteristics of complex systems is their sensitivity to history. This characteristic tells us we can never hope to perfectly test a system outside of production. Production will always be a little different; if not in some subtle architectural or configuration difference, then in the patterns and volume of user behavior and information flow. The chaos monkey operates on the principle that the only way to test production is in production.

One might think that a successful chaos monkey run is one that doesn't cause any visible production problems. If the production systems can survive the chaos monkey, that sounds like a positive result. According to Adrian Cockcroft, though, that's not the case. If you know that failure is lurking within your systems, success involves finding that failure. Netflix considers successful chaos monkey runs to be ones that visibly break the application. Those runs are the ones that show the company how to improve its systems.

Blameless Postmortems

After an outage has come and gone, organizations generally conduct postmortems. A postmortem's ostensible purpose is to identify the root cause for the outage and understand how to prevent it from happening again. Unfortunately, they too often devolve into assignment of blame. Focusing on blame is counterproductive. It discourages the openness necessary to understand the true causes of an outage and thus inhibits learning.

Blameful postmortems also miss the point that complex systems failures often lack a single root cause or even multiple linear causal chains. Searching for someone to blame is thus not only counterproductive but also futile and misguided. Furthermore, if failure is inevitable, as it is in complex systems, then prevention is a fool's errand. Learning is the only viable purpose for a postmortem.

Learning from mistakes requires fearlessness. This imperative leads to the practice of blameless postmortems, pioneered by John Allspaw at Etsy (*http://bit.ly/1E7g3wl*). A blameless postmortem encourages engineers to expose their assumptions, actions, and mistakes without fear of retribution. It encourages participants to share the maximum, rather than the minimum, possible information about their role in the outage. Not only does exposing problems, assumptions, and limitations increase opportunities to improve systems and procedures; it also drives institutional learning by letting everyone witness one another's thought processes.

Finally, blameless postmortems have an important psychological and organizational effect. They treat engineers as intelligent people trying to do the right thing rather than as untrustworthy, potentially defective cogs in the machine. By treating team members with respect, it puts another nail in the coffin of the Taylorist industrial approach to employee management. Complex socio-technical systems require human initiative and creative decision making. Blameless postmortems make a critical contribution to sustaining that spirit within IT organizations.

Operational Transparency

Operational transparency takes blameless postmortems one step further by exposing the details of outages to all interested parties, including customers. One would think that's the last thing service providers would want to do. It would seem to cripple their credibility in their customers' eyes. The fact is, though, that customers know there was an outage. The provider isn't saving themselves any embarrassment by hiding the details. Instead, they are presenting themselves as an organization that believes in honesty, learning, and continuous improvement on its customers' behalf.

We all know from daily life that service doesn't always work perfectly. In fact, we often value a service provider's efforts to fix something for us. If the situation is handled well, we can end up thinking more highly of the service provider after a problem than before. Operational transparency follows this philosophy. It reflects the growing understanding that distributed, digital services are in fact complex and do in fact fail in ways that can neither be foreseen nor prevented.

Designing Businesses for Failure, Operating Them to Learn

The need to accommodate and seek out failure goes beyond the confines of IT. Complexity characterizes the entirety of a digital business and its interaction with customers and the market. In the post-industrial economy, brand management becomes a matter of continuous repair.

No organization can perfectly predict market or customer behavior. Brand failure comes in many flavors. It can take the form of anything from a website outage on Black Friday to a poorly designed new offering. The digital conversational medium helps us better understand and serve customers; it does not, however, guarantee success.

Post-industrial businesses need to take the lessons of complexity to heart on every level. Valuing MTTR over MTBF, for example, makes it possible to minimize the negative brand impact of failure. The power of practices such as Continuous Delivery and microservices transcends purely technical repair. Their greatest benefit is minimizing mean time to repair. A bug corresponds to a unit of customer dissatisfaction; from that perspective, any gap between current and preferred is a bug.

Perfectly continuous customer satisfaction is unachievable. Post-industrial business requires the ability to continuously repair dissatisfaction. One might even claim that the transition from industrialism to post-industrialism involves a shift from MTBF (manufacture millions of identical cars) to MTTR (fluidly respond to disruption).

Continuous repair is a key part of continuous design. As soon as we deliver the things we've designed, they escape our control and lead our customer conversations in unpredictable directions. In order for these conversations to remain useful, we need the ability to continuously repair breakdowns. Digitally infused service depends on IT to conduct conversations and thus to repair breakdowns in them. MTTR-centric IT practices make it possible to repair these breakdowns with minimal latency and therefore minimal customer dissatisfaction.

The need for design-for-fail goes beyond even the user-interface level. Service as a whole has to be able to survive failures of all kinds. If, for example, the customer management system experiences an outage, support representatives still have to answer the phone and provide meaningful information. If a snowstorm makes it impossible for them to get to work, the company still has to help customers with problems. Service design thus needs to consider systemic and human failure just as infrastructure design considers technical failure.

In addition to testing technical design-for-fail solutions, we also need to test human and social ones. Business operations faces similar challenges to technical operations. How will customer support provide meaningful information if the customer management system goes down? If your website fails on the most important shopping day of the year, how will your CEO respond to press inquiries? The only way to know for sure is to simulate potential failure scenarios.

Blamelessness and transparency are becoming business concerns as well as engineering concerns. Nontechnical organizations are beginning to adopt cybernetic approaches to their own work through such practices as Agile marketing. Companies are recognizing the importance of complex-systems structures such as open innovation and platform ecosystems. In order to leverage the power of conversational systems, they also need to transform their attitude toward failure. An Agile marketing campaign, for example, doesn't proceed by striving for perfection. In order to steer its way to success, it needs to analyze and respond to failure blamelessly.

Particularly as social media shifts brand power from the corporation to the customer, consumers have less and less patience for opaque business practices. It's no longer enough for a company to publicly acknowledge a security breach. Customers immediately want to know how and why the breach happened, and what the company is doing to prevent it from happening again. Attempts at information hiding quickly become negative brand moments. Brand repair depends first of all on acknowledging that trust has been broken, and second on communicating the concrete internal activities that can restore trustworthiness.

BEYOND ANALYSIS

Complexity's deepest lesson concerns the limitations of reductionist analysis. There is no way to guarantee the correctness of any possible choice. It's not even possible to guarantee that a reasonable-seeming choice won't make things worse. The best you can do is to try, and to be prepared to continually try again based on the results of each attempt. This approach epitomizes the cybernetic worldview.

If it isn't possible to identify solutions purely through analysis, how can you generate possibilities? Any attempt to manage complex systems needs to incorporate a nonanalytical component. Design thinking offers tremendous insight in this respect. Its humanistic heritage, drawing on philosophy, art, architecture, and literature, provides a powerful counterbalance to IT's engineering legacy.

As much as anything, complex systems force us to embrace new ways of knowing. We need a way to proceed through a world where there is no single root cause, and what causality we find is circular in nature. We need the ability to experiment through disciplined wandering.

The inevitability of failure ironically frees us from the anxiety of trying to "get it right." Instead, we can design for failure, optimize for mean time to repair, and build in feedback loops that bound our wandering. Within those bounds, we can use the designer's mind to choose where to go next. Digital businesses that infuse design thinking the most deeply into their post-industrial control mechanism are best positioned to respond to the challenges of service, infusion, complexity, and disruption.

TRANSFORMING FAILURE INTO SUCCESS

Continuous design, in its guise as continuous repair, functions by continually transforming failure into success. Failure might take the form of a complex-systems outage or a gap between what the customer needs and what the company just produced. It might even manifest as the customer responding to a new feature by saying, "That's wonderful; now how about you make it do this?"

Whatever the cause, if a business stops modifying itself in ways that are meaningful to its customers, it will die, either from stasis or from wandering off into irrelevancy. Complex landscapes change as we move through them, making it impossible to predict the right destination or even the right direction. The only way we can navigate them without foundering on unforeseen rocks or reefs is through never-ending, cybernetic steering.

Conversational IT and the digital business it powers never sit still. They never assume they're done. They never stop listening or responding to what they've heard. The musician Laurie Anderson captured this dynamic in her song "Walking And Falling" (*http://bit.ly/1G98gUD*):

> *You're walking. And you don't always realize it,*
>
> *but you're always falling.*
>
> *With each step you fall forward slightly.*
>
> *And then catch yourself from falling.*
>
> *Over and over, you're falling.*
>
> *And then catching yourself from falling.*
>
> *And this is how you can be walking and falling*

at the same time.

5

The Journey Is the Destination

Post-industrialism demands radical changes in the way businesses understand and conduct themselves. It essentially means turning themselves inside out. Products give way to service. Broadcast gives way to conversation. Hierarchies give way to networks. Scale gives way to speed. Efficiency gives way to adaptability. Robustness gives way to resilience.

Twenty-first-century business confronts IT with the need to undertake an equally radical transformation. IT must turn itself inside out in the same fashion, along similar dimensions. Controlling change gives way to enabling it. Silos give way to collaboration. Systems give way to service. Preventing failure gives way to incorporating it.

Previous chapters have introduced a set of practices that share a cybernetic mindset. They can all be helpful in implementing post-industrial IT. Service design, Agile, DevOps, cloud, and microservices mutually reinforce one another to create a responsive, resilient platform for digital brand conversations.

The key to transformation is not, however, simply to replace an old checklist of best practices with a new one. The central challenge for CIOs is not to make sure their organizations are doing Agile or DevOps, or using cloud or microservices. Ensuring success doesn't just consist of remembering to ask, "Do we hold retrospectives every two weeks?" or "Do we have a chaos monkey?" or "Did we run a game day before last month's release?"

The CIO's most critical role is to lead IT in reimagining its purpose. That purpose is no longer merely to operate information systems. Instead, it is to help businesses continuously change existing situations into preferred ones with and on behalf of their customers.

In order to lead IT's transformation into a medium for conversation through continuous design, CIOs need to guide their organizations in understanding and embracing the point of Agile, DevOps, cloud, design thinking, and whatever other useful practices arise. The point of all of those practices is to enable fluid adaptation to environmental change. That change may come in the form of market disruption, new feature requests, or sudden changes in usage patterns.

IT's new job is to help digital businesses design and operate responsive services, and to empathically engage with customers by designing and operating themselves. The principles of continuous design offer a guide for taking on this daunting task. As part of leading post-industrial IT organizations, and evaluating their success, CIOs can ask themselves, their departments and microservice teams, and their peers questions about whether they are doing the following:

- Designing for service, not just software
- Minimizing latency and maximizing feedback
- Designing for failure and operating to learn
- Using operations as input to design
- Seeking empathy

Designing IT

The principles of continuous design present a method by which IT can fulfill its new role. If that method represents the destination for post-industrial IT, the question remains as to how to chart a path to that destination. How does IT effect such a profound shift in practices, and more importantly, mindset? In particular, how can it avoid turning the transformation effort into yet another massive, failed IT project?

The answer is to treat the journey as the destination. Continuous design applies to IT just as much as it does to the rest of the organization. As part of helping businesses achieve the capacity for continuous design, IT needs to continually design itself. Treating itself as both a source and a target of design rather than as purely an operational organization has a dual benefit. It helps IT improve its usefulness to the business it serves through user-centered responsiveness. At the same time, it gives IT the freedom it needs to experiment, learn, and incrementally improve.

Continuous self-design means applying the principles of continuous design to IT's own systems, procedures, and structure. Designing for service and not just software focuses IT on its purpose rather than on technical implementations. That focus liberates it from attachment to specific solutions. Just as the business as a whole continually evolves its service and the mechanisms for providing that service, so too can IT evolve its capabilities and its inner workings over time.

Continuous self-design implies approaching IT transformation from the perspective of self-steering. Minimizing latency and maximizing feedback are critical to effective steering. On the one hand, pursuing its own internal transformation in an *Agile* fashion lets IT change existing situations into preferred ones more quickly with less waste. On the other hand, it requires a dramatically different approach to process improvement and system implementation. It necessitates letting go of the traditional, waterfall model of evaluating, choosing, and then mandating new methodologies and their supporting systems in favor of adoption through small-scale experimentation followed by adaptive propagation.

Designing for failure and operating to learn frees IT from the illusion that any set of methodologies or systems will perfectly solve the transformation problem. It also recognizes the reality that even the problem itself can't be perfectly understood in advance. This approach lets IT approach any given incremental transformation effort with less anxiety and, ironically, with greater chance of success. The assumption that simply adopting Agile, for example, will turn IT into an ideal servant for the business creates paralysis in the face of imperfection. A continuous design perspective, by contrast, views implementation problems as information, rather than as evidence Agile doesn't work and should be abandoned.

In order to complete the continuous design circuit and allow success to emerge from conversation, information about failure needs someplace constructive to go. Using operations as input to design means taking feedback seriously, valuing and acting on it, and allowing IT systems, processes, structures, and strategies to flex and change in response to it. Doing so creates a culture of flexibility, adaptability, openness, and incremental improvement. It lets everyone in IT and beyond see that the transformation process is never-ending. It frees everyone in IT to experiment, to report the results of their experiments, and to respond to others' results without fear.

Empathy is the starting point as well as the result of the entire process. Approaching IT transformation as continuous design emphasizes its benefit to others over any specific destination it might pursue. It evaluates IT at any point in time in terms of its usefulness, and continually strives to change its current state of usefulness into a preferred one. Minimizing latency, maximizing feedback, learning from failure, and using operations as input to design all reinforce the primacy of listening to and learning from external customers and internal users.

Equally as important, continuous design both fosters and relies on mutual empathy among teams and individuals within IT. By shifting IT's focus from stability and correctness to exploration and adaptation, it frees the various parts of IT to listen to and learn from one another rather than competing or defending themselves, or trying to hold one another at bay out of fear of error. On the other hand, continuous design doesn't work unless everyone embraces this shift in mindset. Individuals have to go through the same transformation as the organization and process in which they're participating. As much as anything, the CIO's job is to facilitate that internal transformation.

IT Transformation as Urban Architecture

As if the post-industrial challenge weren't sufficiently daunting, IT organizations also have to figure out how to transform themselves while simultaneously maintaining business-critical legacy systems. How does one replace the engine while the plane is in flight? To answer this question, we can learn from the lessons of urban architecture. In particular, the story of the Leadenhall Building in London is instructive and inspiring.

The Leadenhall Building is a recently completed skyscraper in central London (Figure 5-1). It's also known as "the Cheese Grater" because of its triangular shape. That shape, along with many of the building's design and construction constraints, reflects its location in the midst of a major, 2,000-year-old city. By law, the building is prohibited from obscuring the view toward Saint Paul's Cathedral. Its triangular shape allows it to have the desired number of floors without violating the city ordinance. Unfortunately, the shape creates its own problems. It means that the highest, most profitable floors have the least square footage.

Figure 5-1. Leadenhall Building and The Royal Exchange (http://bit.ly/1ELlQZ4) by Ronnie Macdonald (http://bit.ly/1ELlRw9) (licensed under Creative Commons (https://creativecom mons.org/licenses/by/2.0/))

Skyscrapers normally put service systems (elevators, bathrooms, plumbing, and electrical) in a central corridor that runs up the middle of the building. To maximize office space on the upper floors, the Leadenhall's architects placed its service systems to one side. That solution in turn caused balance problems for the building. The architects therefore designed a revolutionary shim system that would let the construction crew shift the building on its foundation after construction was finished.

The construction team faced other problems because of the building's location within the city. The building site was very small, without any wide access streets. These limitations made the normal practice of mixing concrete and assembling steel beams on site impractical. Instead, the design team completely reimagined the construction process. They designed the building to be made of relatively small prefab parts that could be built outside the city, and then trucked in and incrementally assembled on site.

Throughout the project, the Leadenhall's architects treated daunting constraints as creative challenges. Rather than insisting on applying traditional methods or giving up and declaring the project impossible, they integrated the constraints into the design process. They engaged in conversation with the environment rather than trying to impose their will on it. As a result, they designed not just a skyscraper but an entirely new approach to skyscraper design and construction.

The story of the Leadenhall offers several lessons for IT and legacy, whether it be systems, process, or structure:

- Transform implementation obstacles into design constraints
- Don't believe success is binary
- Challenge your own assumptions
- Reframe the problem
- Innovate where you can

Continuous Refactoring

The truth is that IT never completely frees itself from legacy processes, tools, or systems. As soon as an organization transforms the current into the preferred, the preferred becomes the current. What solves today's problem doesn't solve tomorrow's problem. Virtualization was an improvement over physical servers. Containers are an improvement over virtual machines. Containers introduce their own problems and thus new opportunities for design.

The same truth holds for the practices IT uses to manage and operate itself. Continuous design liberates IT from the impossible pursuit of perfection, either in what it produces or in how it produces it. It turns the evaluation of new methodologies into a cybernetic conversation. Whether it be ITIL, DevOps, or whatever comes after both, the adoption process becomes one of continual questioning:

- What does this method mean to us?

- How do we need to adapt it to our environment?

- How does adopting it change us?

- What new questions do those changes present?

So-called best practices culled from industry success stories become signposts rather than destinations. Etsy and Netflix, for example, both practice Continuous Delivery. However, each company takes a somewhat different approach. The differences between their respective implementations reflect their divergent businesses, cultures, and histories. Every IT organization needs to actively internalize new methods (including, of course, continuous design itself) rather than passively imposing them on itself.

Providing a medium through which digital businesses can continually design themselves is IT's new mandate. To fulfill that mandate, IT has to shed its own industrial mindset. It needs to embrace continuous refactoring—changing current internal situations into preferred ones—as its fundamental operating model. As organizations evolve in response to customer needs and market demands, the things they need from IT evolve as well. Post-industrial IT thus represents a double transformation: empowering itself to self-steer in service of empowering the organizations it serves to do the same.

Effecting such a pervasive transformation requires IT to take a cybernetic perspective on all of its systems and activities at all levels from itself as a whole to each team and individual and their relationships to one another. It returns the "I" in CIO to its original meaning from the dawn of computing. The post-industrial CIO's job description becomes nothing less than the initiation, encouragement, guidance, and rewarding of receptive communication as the daily way of being for the entire organization.

Continuous Quality

6

A New Definition of Quality

In order to navigate the challenges of the post-industrial economy, twenty-first-century businesses need a new, cybernetic approach to control. The new business imperative in turn demands a transformation on IT's part from an efficiency aid to a medium for empathic digital conversations. To accomplish this transformation, IT must radically redefine its understanding of quality.

IT's new role involves doing more than just enabling a company's ability to conduct conversations with its customers. These conversations must be useful. They must lead to brand enhancement rather than degradation. They must help the organization dynamically maintain its viability. Merely unifying design and operations isn't sufficient; continuous design only succeeds when it helps the organization steer in desired directions.

In order to be effective and not just fast, the digital conversational medium must incorporate a mechanism for validating the usefulness of conversations and their impact on brand quality. Cybernetics supports self-validating processes by building detection of error into its basic model of control. Defining design as "changing existing situations into preferred ones" implies a mechanism for determining whether changes are in fact preferred.

Within engineering organizations in general and software development in particular, Quality Assurance (QA) traditionally plays the role of detecting error and validating desirability. Before one can address how to detect gaps between actual and desired, however, one must start by defining quality. Continuously identifying and correcting the gap between actual and desired requires a shared understanding of what in fact *desired* means.

Defining software quality in industrial terms is straightforward. It takes the form of a simple question: "Does the software meet the spec?" When I was asked how I defined quality during my interview for the QA job at Apple all those years ago, the Agile Manifesto was still 15 years in the future. Software development followed a waterfall approach. Software projects managed requirements, design, implementation, testing, and delivery as separate phases. One phase didn't begin until the previous one was complete.

Not only did software engineering keep phases separate from one another; it kept teams separate as well. QA and development intentionally maintained an arms-length, adversarial relationship. Software production followed an assembly line approach, with work moving linearly between phases until it left the "factory" for delivery to the customer.

Engineering teams believed in the importance of getting the requirements right up front. They believed that effort spent on adjusting the product in response to changing understanding was wasted effort. Properly understanding, documenting, and communicating requirements prior to development was key to maximizing the efficiency of the software production process.

QA didn't view its job in terms of providing feedback that development could use to improve its understanding of requirements. Instead, QA saw its job as validating whether development had implemented those requirements correctly. While never feasible, software that went from development to testing with zero bugs was always the holy grail. Project managers considered the number of bugs to be inversely proportional to software quality.

The old definition of quality reflected software's product nature. Software companies built software; their customers took responsibility for operating it. Product release notes included instructions on how to install the software, what computational resources it required to run, and other specifications. Some of us may even be old enough to remember the days when software arrived in a box in the mail. It was a physical thing. You bought it, opened the package when it arrived, and stored it on a shelf in your office.

Software as product implied it was finished when it was delivered to customers. We used words like *Final* and *Golden Master*. Among other things, the focus on getting requirements right up front reflected the assumption that requirements could ever be *done*. Although it was virtual, and malleable in nature, vendors and customers both treated software as if it were a physical object like a car. Even if you leased a new car every three years, when you signed the lease, you were stuck with that model until your lease was up. Only then could you upgrade. Similarly, software companies released new versions relatively infrequently.

The product nature of software reinforced long release cycles. Customers bore the brunt of adopting change. If a new version required more memory or an operating system update, the onus was on the customer to implement the necessary upgrades. Customers thus served as a brake for change. They contributed to the treatment of software development as a discontinuous rather than continuous process.

Post-Industrial Quality

Now, though, we need a new definition of quality that matches the challenges and realities of the post-industrial era; one that reflects the transformation of software from product to service. It has to account for the ways in which complexity changes the meaning of success and failure. It needs to address the impact of digital infusion and disruption on the role of software in business.

The principles of continuous design define a methodology for implementing post-industrial IT. Continuous-design organizations still, though, need a set of criteria by which to evaluate their effectiveness in:

- Designing for service, not just software
- Minimizing latency and maximizing feedback
- Designing for failure and operating to learn
- Using operations as input to design
- Seeking empathy

The criteria by which one would validate continuous design–based service quality would include questions such as:

- Does the service helps customers accomplish their jobs-to-be-done?
- Is the service resilient and adaptive?
- Does the service help the business self-steer?

Does the service help customers accomplish their jobs-to-be-done? This question addresses functional requirements in service-centric terms. It validates whether the service is functioning as it should. From a service-oriented perspective, proper functioning means helping customers accomplish their goals.

This question provides a lens through which to evaluate the entire service organization's alignment within itself and with its customers. It turns that lens on service as a whole, not just the computerized component. It helps improve quality by pushing the organization to consider human elements of service such as documentation, training, and support.

Is the service resilient and adaptive? This question addresses nonfunctional requirements in service-centric terms. It views nonfunctional requirements and the infrastructure that supports them from the perspectives of cybernetics and complex systems. It helps improve quality by encouraging the organization to evaluate itself from a complex-systems perspective on success and failure, robustness and fragility.

Does the service help the business self-steer? Businesses self-steer through service. Steering requires internal as well as external coupling. Microservices maintain their viability by adapting to the needs of other organizational components. Organizations maintain their viability by propagating the needs of customers and the markets through those internal relationships. The extent to which each part of an organization can hear and respond to feedback, and continuously redesign itself in tune with the company as a whole, is the deepest and most subtle measure of digital service quality.

Jobs-To-Be-Done

Quality means more than just building something properly. It also means building the proper thing. The best-written software does no one any good if it doesn't meet real user needs. The industrial "Does the software meet the spec?" definition of quality separates these two concerns. It assumes that the requirements process has correctly captured customers' needs. It leaves that process out of the cybernetic loop of learning and adaptation.

By explicitly incorporating fitness-for-purpose, the new definition of quality reunifies building-it-right with building-the-right-thing. It focuses the organization beyond just the service itself and toward the service's purpose. Even more importantly, it helps the organization look entirely beyond itself to the customer's purpose.

Clay Christensen, who developed the theory of disruptive innovation, also coined the term *job-to-be-done* to capture the importance of addressing customer purpose. He used it to refer to a user-centered approach to marketing as an alternative to traditional market segmentation. He claimed that people don't buy particular products because they're 18–34 years old or because they live in a suburb, or based on any other demographic information. Instead, they *hire* products to help them accomplish specific *jobs.*

Christensen illustrated his thesis with the example of commuters buying milkshakes. Why do fast-food restaurants make their milkshakes so thick and hard to drink through a straw? They discovered that people often bought milkshakes to drink on long, boring commutes. They wanted something with which to occupy themselves over the course of an hour or two of driving.

The job for which fast-food customers "hired" milkshakes wasn't immediate refreshment or nutrition, but rather to help themselves pass the time. When seen from this perspective, the milkshake's thickness and the slowness with which it softens helps rather than hinders. Christensen had worked with a client who'd tried making its milkshakes thinner, without positive results. Sales counterintuitively went up when the client made its shakes even thicker than they'd originally been.

Christensen believed that you can more accurately target a product by aligning it with customers' goals and desired outcomes instead of your view of the market. Centering the definition of quality around jobs-to-be-done drives services to empathize with their customers rather than trying to manipulate them into buying what the business wants to sell. This approach aligns with the postindustrial, conversational nature of service marketing and delivery. It makes it possible to incorporate requirements themselves into the cybernetic learning loop. A service organization can use questions about the customer's job-to-be-done to help validate the true usefulness of proposed requirements:

- What are the customers' jobs-to-be-done?
- What help do they need to accomplish their jobs?
- What does/should our service do to help them accomplish their jobs?
- How does a given function of our service help customers accomplish their jobs?

Service-Dominant Logic

The jobs-to-be-done model reflects a deeper truth about the nature of value in the context of service. Vendors don't deliver value to customers. Instead, vendors and customers co-create it together. We can see this truth reflected in the way the jobs-to-be-done model uses the language of *hiring to help.*

In order to truly achieve post-industrial quality, software service organizations need to deeply understand and internalize the concept of co-creation. Value co-creation through service exchange is a central element of the theory of *service-dominant logic*, which was developed by Stephen Vargo and Robert Lusch.[1]

Vargo and Lusch explain service-dominant logic by contrasting it with the existing economic paradigm, which they call *goods-dominant logic*. Good-dominant logic is the primary lens through which economists have viewed business for the past several centuries. It puts the creation and distribution of goods, or products, at the center of economic activity. According to goods-dominant logic, producers create value in the form of tangible and intangible objects. They then exchange those objects with customers in return for money.

Goods-dominant logic describes an asymmetric relationship between producers and consumers. The word *consumer* implies a passive recipient rather than an active participant. Artists are producers; audiences are consumers. Even worse, goods-dominant logic defines consumption as *value destruction*. From a product-centric perspective, that disdainful-sounding moniker actually makes sense. If you buy a candy bar and eat it, you've destroyed the value contained within it. No one else can access the enjoyment or refreshment that the candy bar afforded you.

Service-dominant logic places service exchange at the center of economic activity. Goods are merely *appliances* that enable service exchange to happen. From a goods-dominant perspective, when I buy a car, I'm exchanging money for the value contained within the car. From a service-dominant perspective, on the other hand, I'm paying the dealer—and indirectly the manufacturer—to operate businesses that design, build, market, sell, and service cars on my behalf. I'm paying them to have a car available that meets my needs when I'm ready to buy one. I'm also paying them to apply the necessary effort and expertise to make sure the car is safe, gets reasonable gas mileage, handles well, and so on.

Service-dominant logic transforms the asymmetrical producer–consumer relationship into a symmetrical relationship between actors. In order to have enough money buy the car, I have to exercise my own effort and expertise. I need to have a job and get up every morning and go to work. I need to succeed at my job in order to have enough money on hand when the time comes to purchase

---

1 Vargo and Lush introduced the term in 2004 in an article titled "Evolving to a New Dominant Logic for Marketing." They explore their theory in more detail in the 2014 book, *Service-Dominant Logic: Premises, Perspectives, Possibilities*.

the car. Buying it lets me use my money to help the dealership's owner meet his own needs. Service-dominant logic thus defines all economic activity as "the exchange of service for service."

The term *co-creation* refers to the fact that actors create value together while exchanging service. The car is only valuable to me because it lets me drive to work and to the grocery story. It was only available for me to buy, though, because the dealer ran a successful business. The money I pay for my car lets the dealer continue to do so.

Co-creation means that value exists in the relationship and in the exchange, not in the thing being exchanged. Each party gains value when an exchange helps in attaining desired outcomes. We can define quality as a measure of success at creating value. A goods-dominant approach to quality measures how well the object being produced fulfills the vision of what it should be. In other words, does the software meet the spec? A service-dominant approach to quality measures how well the service helps customers accomplish their jobs-to-be-done.

SERVICE-DOMINANT LOGIC AND CYBERNETICS

Service-dominant logic is highly compatible with the cybernetic mindset. That mindset views individuals and their environments as inseparable parts of a dynamic relationship. Service-dominant logic rests on the same conceptual foundation. Buyers and sellers co-create value with one another through service exchange as part of their respective autopoiesis. After driving my car to and from work for 10 years, I wear it out and need a new one. Having driven to work all those years, I have enough money to be able to replace it. Being able to afford a new car means I can continue driving to work and making enough money to pay my mortgage and buy groceries.

The conversational nature of service-dominant economic activity means that all value is not only co-created but also mutually beneficial. Buying a new car doesn't destroy the car's value. The dealership doesn't lose its value by transferring it to me in the form of a sale. The dealership gets as much value from the exchange as I do. My money allows the dealership to continue in the business of selling cars. Making a profit allows the dealership's owner to provide for his family and employees. It enables the dealership's own autopoiesis.

Service exchange corresponds to the structural coupling that takes place between autopoietic systems. It allows buyer and seller to self-steer by having useful conversations with each other. Service exchange defines a dynamic, mutually beneficial relationship that allows all parties to adaptively maintain their viability.

SERVICE-DOMINANT LOGIC AND SYSTEMS THINKING

This last point illustrates another important tenet of service-dominant logic. All participants in service exchange are *service integrators*. In order to co-create value with one another, the participants must in turn exchange service with other entities. I can't buy a car without a financing company. The dealer can't sell the car to me without a manufacturer. They likely also need to lease the land and the building from a real estate company. The real estate company has its own relationship with a bank. The manufacturer depends on suppliers, and so on.

Understanding service quality requires the ability to understand systems thinking. Just as the thermostat holistically implies larger surrounding contexts, so too does the concept of value as co-creative exchange between service integrators. The quality of your service depends on its ability to fit within a larger system of service exchanges. The sole purpose of any activity in which your organization engages or any thing it produces is to enable service integration and exchange.

SERVICE-DOMINANT LOGIC AND CLOUD COMPUTING

We can illustrate the impact of service-dominant logic on software through the example of cloud computing. If there were any doubts about the transformation of computing resources from product to service, terms like Infrastructure-as-a-Service (IaaS) and Software-as-a-Service (SaaS) should dispel them. IaaS isn't really about acquiring instances of Linux servers. In reality, it's actually about helping SaaS companies cost-effectively operate scalable, resilient web applications.

Amazon Web Services (AWS) understands this fact better than any of its competitors. AWS doesn't just offer servers-on-demand. It offers an entire suite of virtual, on-demand application building blocks: everything from servers to load balancers to storage appliances to databases.

The latest Gartner IaaS Magic Quadrant (MQ) puts AWS so far ahead of everyone else in the IaaS space that it had to enlarge the scale to show AWS on the same diagram as its competitors. AWS's accelerating advantage reflects its focus on helping customers accomplish their jobs-to-be-done (building and operating applications) rather than on its own goods (virtual servers).

Amazon can't accomplish this feat on its own. The company doesn't write its own operating systems or databases. It integrates open source systems such as Linux and MySQL, and Microsoft products such as Windows Server and SQL Server, into its offering. Amazon's value proposition consists of making it easier for customers to use these systems.

Cloud computing thus defines a multilevel process of service integration. A small business might hire an invoicing SaaS to help manage its finances without having to run its own finance software. The SaaS company in turn hires AWS to help operate its invoicing service without having to run its own data center. AWS completes the equation by hiring Microsoft to help provide on-demand infrastructure without having to write its own operating system.

The Customer Journey

Value co-creation through service exchange differs from goods delivery in another important way. Customers experience service via journeys over time across multiple touchpoints. Imagine that you're taking your spouse out to dinner on your anniversary. Your experience begins when you go online to find a restaurant. The restaurant's online menu is its first interaction with you. Next, you call to make a reservation. Then you arrive and have to find a place to park. Perhaps the restaurant has a dedicated parking lot; maybe it has valet parking; or maybe you have to fend for yourself and find on-street parking.

These touchpoints all influence your state of mind when you walk into the restaurant. By the time your dinner is over and you're on your way home, you've interacted with the maître d', the waiter, the menu, and indirectly with the kitchen staff that actually prepares the meal. You may have had a drink at the bar while you waited for a table. The act of eating is just one element of the overall experience. Even after your meal is over, further touchpoints remain. The speed and accuracy with which the restaurant prepares the check and processes your payment, for example, leaves a final positive or negative impression.

Customers judge services by the entirety of their experience. Imagine you've tried a newly opened restaurant. A friend asks you how it was. You answer, "The food was great but the service was terrible." Based on your experience, your friend is probably a lot less likely to try the restaurant for herself.

The experiential nature of service applies to software just as much as any other kind of service. Using an IaaS console to provision an on-demand virtual server is analogous to eating the food in the restaurant. In order to reach that touchpoint, you've had to navigate the front end of your journey. To start with, you had to understand what the service is and how it works. Interacting with the provider's website, white papers, and marketing staff is actually the beginning of the service relationship.

Once you agree to extend that relationship by becoming a customer, you need to go through an onboarding and training process. Only then are you ready to provision servers. Once you start using the service, you may need help. The service may have an outage or a security breach.

Supposedly "ancillary" functions such as onboarding, training, and help are all integral components of the service experience. They are all required in order to co-create value. If running an application on the IaaS provider's platform is your job-to-be-done, and you're having trouble, you need help. Without that help, your ability to accomplish your job-to-be-done is compromised.

SERVICE OUTAGES AS SERVICE EXPERIENCES

Software service providers often impair their ability to deliver quality in customers' eyes by not attending to the entire experience. Outage management is a great example. A customer who uses a software service is effectively outsourcing part of its IT organization. That outsourced component must function as a part of the overall team.

Imagine a scenario wherein a small business uses an invoicing service and that service has an outage. The CFO has the following dialogue with the company's IT director:

> CFO: "Why can't we generate invoices?"
>
> IT: "Because our invoicing service is having an outage."
>
> CFO: "Why?"
>
> IT: "I don't know."
>
> CFO: "When is it liable to be fixed?"
>
> IT: "I don't know."
>
> CFO: "What's being done to prevent a similar outage in the future?"
>
> IT: "I don't know."

One can easily imagine that neither the CFO nor the IT director will consider this dialogue satisfactory. A happier scenario can be seen in the way Zappos handled a security breach in 2012. The company notified customers immediately. It took full responsibility for the breach without making any excuses. It communicated clearly, fully, and directly. The visual and written style of the communications was completely in keeping with its customer-centered corporate vision. As a result, instead of seeing its brand damaged, Zappos actually reinforced it. The company understood that breach management and notification was a customer service touchpoint like any other.

A New Definition of QA

Control is the universal underlying purpose that drives business activity. Quality is a measure of control. When a company creates a product or service that functions as intended, and through that functionality succeeds in helping customers accomplish their goals, it demonstrates control over its own development process. The ultimate measure of control is an organization's ability to perpetuate its own autopoiesis through self-steering; in other words, to leverage the interaction of its component parts in order to stay in business by continually adapting to the market.

Quality assurance is the quintessential cybernetic process. It provides the feedback necessary to narrow the gap between desired and actual behavior. Its very presence demonstrates some recognition that achieving quality without course correction is infeasible. Over the years, there have been numerous efforts to engineer defects out of the software development process. Some have taken a procedural approach through activities such as rigorous requirements analysis. Others, such as language designers who tried to create programming languages that made it possible to write provably correct programs, have taken a linguistic approach. None of them, however, has succeeded.

QA thus serves a dual purpose: it gives organizations visibility into their level of control while simultaneously helping them improve it. In true cybernetic fashion, it becomes part of the control process. By providing information about the gap between actual and desired, it enables steering. QA serves as the mechanism that allows an organization to listen to itself. As half of the self-steering conversation, it plays an indispensable role in an organization's autopoiesis.

In order to measure quality given its new, post-industrial definition, QA must unify the perspectives of cybernetics and service-dominant logic. On one level, it measures how well a service helps customers accomplish their jobs-to-be-done. On another level, it measures how well the service organization responds both internally and externally to conversations; in other words, how useful its internal and external conversations are.

Post-industrial quality resembles a physical property that is at once particle and wave. The principles of continuous design try to capture this dual nature. Post-industrial QA's job is to validate an organization's effectiveness in continuously changing existing situations into preferred ones. To accomplish this job, QA must simultaneously measure the state of a service and the process by which that state changes over time.

ASSURING EMPATHY

Post-industrial quality is a measure of how well all parts of a service provider's organization understand and work with each other and with customers. Service quality is a reflection of internal seamlessness from top to bottom and from front to back. The holistic, systemic, experiential nature of service means that everyone in the organization must understand the entire customer journey and their place in it. They must see themselves as participating in value co-creation by helping customers accomplish their jobs-to-be-done.

Empathy is critical to all of these capabilities both as their drivers and their outcomes. At its core, therefore, post-industrial QA is about measuring a service provider's ability to empathize with itself and its customers. Its greatest value lies in helping an organization see the places and ways it needs to empathize, and continually improve its ability to do so.

7

The Four Dimensions of Digital Service

A company's ability to co-create value through digital service manifests along multiple dimensions. Customers judge the quality of their experience across all dimensions inseparably. Just as with a restaurant that offers great food but lousy service, a software service organization will lose customers if it falls short in any of these areas. It needs to approach them as truly being different dimensions of a unified whole.

Traditionally, formal QA activities have focused primarily on validating functional correctness. In order to ensure their ability to view themselves and their customers more holistically, digital service organizations need to take a more holistic approach to QA. They need to reimagine QA as a feedback mechanism for service, not just software, in the process transforming it into a proxy that helps the entire organization empathize with its customers.

As part of this transformation, they must shift their focus away from the software, and toward the customer. From that perspective, we can define four dimensions along which customers judge digital service quality:

Outcomes
> The extent to which software functionality enables desirable customer outcomes

Access
> The extent to which operations enables customer access to software functionality in order to accomplish those outcomes

Coherency
> The extent to which a service enables coherent customer journeys

Continuity

> The extent to which a service continuously redesigns itself in response to evolving customer needs and desires in the other three dimensions

Together, these four dimensions define the criteria that characterize high-quality, continuously designed digital services. This chapter examines them in turn, explaining the specific ways each one provides a basis for service-centric QA practices. It then describes how they come together into a mechanism for validating the effectiveness of continuous design as a coherent whole.

Outcomes

In order to accomplish a job-to-be-done, a customer needs to complete a variety of tasks. Quality of outcomes measures the extent to which the service helps the user successfully engage in those tasks. Consider, for example, an online invoicing service: as an independent consultant using this service, what is my job-to-be-done? To put it simply, I want to get paid on time. In order to accomplish that goal, I need to do things like track my time and generate and submit invoices. Those tasks generate requirements for ancillary tasks such as setting up new clients and projects, entering email addresses, and so on.

QA in the outcomes dimension has three levels. The first, and most familiar, involves validating whether a given feature works properly. When I generate an invoice from the current week's time, for example, does the invoice contain the right number of hours? Is the calculated total dollar amount correct? Does the invoicing feature still work properly if I have a large number of subcontractors who have entered time for a large number of clients? Does it prevent me from sending the client an invoice without a required PO?

The second level involves validating usability. Is the interface to a feature or the flow between features clumsy or convenient? Is it easy or confusing to figure out? Does the arrangement of functions match my natural workflow or does it increase cognitive load by making me adjust my mindset to match the software? If I can't use a feature, it doesn't matter whether it does what I want it to do.

The third level of QA in the outcomes dimension is the most subtle and also the most important. It transcends proper or smooth functioning. How well does a set of features and their interface demonstrate the company's ability to understand and empathize with the user? In other words, how well do the functional tasks actually map to what I'm trying to accomplish? If getting paid on time is my job-to-be-done, does the service help me do that or does it just let me execute random billing-related tasks? Does the company even understand what my jobs-to-be-done or desired outcomes are?

Thinking in terms of jobs-to-be-done helps improve quality of outcomes. As part of getting paid on time, I need to track invoice aging. I also need to stay focused on doing work in order to have something for which to bill. I can't afford to always be checking the status of my outstanding invoices. I shouldn't have to remind myself that I have some that need attention. A good service will remind me to pester clients whose invoices are late. Even better, it will let me know that I have some invoices that are nearing their payment deadlines. Best of all, it will let me configure reminders to reflect the flow of my business and my personal working style.

Getting paid on time in turn is part of a larger job-to-be-done. My ultimate goal is to run a profitable business. To do that, I need to be able to integrate my invoicing data with other financial data. I'd probably like to integrate my invoicing service with my general financial management software. Measuring quality from the perspective of jobs-to-be-done helps uncover opportunities to improve alignment between a service and its customers' needs by seeing the larger context around those needs.

The third level of QA necessitates going beyond just validating software based on what it is. It requires the ability to put oneself in the customer's shoes. It involves seeing the world and the service from the customer's perspective. To play this critical role, QA has to become a proxy for the customer. The QA team needs to empathize deeply with the customer, and reflect that empathy back to the development organization as part of the feedback it provides.

Access

Software-as-a-Service shifts the onus for operating applications from customers back to vendors. No matter how well a company understands my need to get paid on time; no matter how well it maps that understanding to feature sets; no matter how beautifully, elegantly, and practically it designs the interface to those features—none of it does me any good if the service isn't available when I need to use it. Quality of access is the dimension that measures customers' ability to access a service when and how they need to in order to accomplish their desired outcomes.

Software-as-a-Service brings operations up out of the basement and makes it part of the explicit value proposition for which customers pay. If you think back to the early days of SaaS, you might remember that as part of their efforts to educate prospective customers about their value, vendors mentioned things like redundant data centers, offsite backup, industrial-grade security, and instant access to application upgrades as features. In other words, they spent significant marketing time and energy on operations.

Access has numerous characteristics, all of which need to be addressed as part of the operability dimension of quality. These characteristics include:

Availability

> Is the service up and running at the end of the month when I need to bill my clients?

Scalability

> Does the service perform adequately, whether I'm the only user on the system or one of a million?

Consistency

> Does the service ensure my data doesn't get corrupted or lost? Can it restore lost data?

Security

> Does the service prevent hackers from stealing my personal information, or using the service to hack me?

Visibility

> Does the service provide operational information how and when I need it?

Access testing stresses the capabilities and toolsets of most QA practices that have traditionally focused on functional quality. Quality of access requirements force developers and testers alike to think at a deeper system level. They need the ability to understand—and to understand how to validate—failure modes involving fundamental computer science topics such as distributed systems, resource scaling, and security.

They also need to be able to understand complicated hardware and software architectures. How do interactions between compute, storage, and networking, for example, impact potential failure scenarios? The same question applies to interactions between applications and middleware, and between applications participating in a service-oriented architecture.

OPERATIONS AS INTERNAL CUSTOMERS

Quality of access involves more than just the robustness of a service infrastructure. It also includes the ease with which human operators can manage the service. In order to make sure disks don't fill up, memory doesn't overflow, and networks don't become congested, operations staff need tools that provide the visibility they need, when and how they need it, and that let them efficiently and effectively take appropriate action.

Operations is an often-overlooked internal customer for a digital service. This omission applies equally to business as well as technical operations. In order to answer customers' questions or make changes on their behalf, for example, customer support staff need access to their own tools that enable visibility and effective action.

Viewing technical and business operations as service customers is a critical part of the value QA can provide. Customer satisfaction depends as much on quality of access as it does on quality of outcomes. Quality of access depends not just on technical infrastructure but also on humans' ability to interact with that infrastructure. The design of a digital service must account for all its customers, both internal and external. QA needs to incorporate internal customers into its measurement of design quality.

This requirement further stresses traditional QA capabilities. In addition to empathizing with customers, QA needs to empathize with operations. To do so, it needs to understand their jobs-to-be-done and the special challenges they face in doing those jobs.

FAILURE AND OPERABILITY

The complex nature of software services makes outages inevitable. Managing outages becomes an integral part of operating a service. Outage management requires more than just diagnosing and fixing technical problems. Quality of access measures more than just how quickly an operations team can repair service. It also measures how well a service communicates with its customers during and after outages.

In 2012, Amazon Web Services suffered a major, 20-hour outage. During the first part of the outage, Amazon communicated very minimally. When taken to task for not being more transparent, the company explained that it had assumed customers would prefer for more time to be spent fixing the problem than communicating about it. As the outage continued and Amazon began communicating more fully, customers' frustration levels counterintuitively decreased. The tenor of Twitter discussions changed over the course of the outage from annoyance to rooting for the AWS operations team.

Transparency on the part of the vendor generated empathy on the part of customers. From the customers' perspective, the quality of the outage-management process increased. Measuring quality at this level must be part of QA's job.

Complex systems' failure-prone nature also necessitates the ability to fail gracefully on multiple levels. The design-for-fail approach is needed from the infrastructure layer all the way up to the UX layer and beyond. The outcomes and access dimensions interact at this point. When considering how a service should act and look, designers need to consider failure modes in addition to normal operational modes. QA needs to do the same; it must validate the quality of graceful failure mechanisms at all layers.

Coherency

Coherency is the dimension of quality that measures customers' ability to engage with a service throughout their journey. Users judge service quality by the entirety of that journey. If, for example, the new-user onboarding process is too complicated or confusing, it might drive customers away before they ever use the service "for real." They might never find out just how well designed the primary service UI is because they never get that far.

Coherency involves more than just doing the right thing at each point in time. Just as an application's user interface must "hang together" as a whole across all its parts, so too a service must hang together across time and touch-points. To achieve this goal, a service organization has to think in terms of systems and dynamic processes. The very use of the word *journey* implies a coherent progression through time and space.

In the process of using a service, I continually make transitions. The job of the service provider is to help me make them. The first is the transition from noncustomer to customer. My journey starts before I even sign up. My ability to understand what a service is, how it helps me, and what I need to do to use it contributes to measuring its overall quality.

The onboarding process is one of the most important yet neglected customer journey transitions. It sets the stage for the rest of my experience. Onboarding includes not only service provisioning and possibly data import but also learning. It reaches completion when I understand the service well enough to use it comfortably. If a service doesn't do a good job of helping me get onboard, I will struggle through the rest of my journey.

Even after I've successfully adopted a service, I continue to go through transitions as part of my daily interaction with it. If I have a problem or uncover a bug, I transition into customer support mode. If I use the service as the leader of a team, I likely move back and forth between ordinary usage and administration. At some point, I will inevitably transition into (and hopefully back out of) outage mode.

Eventually I might decide to leave the service. Many SaaS providers take the department-store approach and make it as hard as possible to leave. They need to remember, though, that departure represents a final opportunity to leave a good or bad impression. Customers cancel their service for a reason: perhaps they're dissatisfied and leaving in favor of a competitor or maybe they simply no longer need the service. In either case, the service provider has a precious opportunity to use this transition as feedback for its own self-steering.

Addressing quality in the coherency dimension pushes companies to fully align themselves with their customers' jobs-to-be-done. Customers devote time and expend energy on transitions because they need to accomplish things. I switch from usage mode to administration mode because I've hired a new employee or terminated an existing one. In addition to needing each service function to work well, I also need the transitions between them to be as seamless as possible.

A service exists to help me accomplish my goals in some domain. In order to fulfill my job-to-be-done, I need to engage in a suite of activities. To oversee a team project, for example, I need to add new members, edit my own data, review others' changes, and so on. A high-quality service always points back to my job-to-be-done. It lets me follow the flow of activities through my journey in a natural way rather than forcing me to adapt my conceptual model to the one preferred by the service designer.

QA plays a critical role with respect to coherency by validating the user-centeredness of a service's design. Teams within service organizations naturally focus on specific touchpoints, sometimes losing sight of the connections between them. In addition, digital service companies tend to emphasize core usage touchpoints to the detriment of other aspects of the overall customer journey. Onboarding, help, and administration functions too often provide unfortunate contrasts to an otherwise high-quality user experience.

QA in the coherency dimension provides a counterbalance by reminding individual teams of the customer journey that surrounds them. It fulfills this function by continually asking questions such as:

- How do users figure out how to use a given feature?
- How do they configure it?
- How do they get help with it?
- How do they they transition between it and adjacent features?
- How do they find out about and understand changes to it?
- What do they do if it's not working properly?
- Where are there discontinuities or entirely missing elements within the overall journey?

EMPLOYEE AND CUSTOMER JOURNEYS

Service-dominant logic implies that quality depends as much on company behavior as on customer behavior. Not only do users need to be able to map their jobs-to-be-done to service journeys; employees also need to be able to do the same thing. Any internal struggles and discontinuities will manifest as customer frustration.

Customer support illustrates the need for synchronization between internal and external coherency. When a user calls the support line, the staff member who answers the call needs to quickly find the right customer record. The staff member then needs to associate that record with useful information about the user's account and history. We all know how frustrating it is to call for help and have to repeatedly explain who we are and what kind of account we have before we can ask the question that motivated our call in the first place.

To provide good support, employees need to understand not only who the customer is but also where he is in his journey. Without understanding the flow of transitions, the employee will struggle to fully comprehend the customer's problem. If I call support because I can't figure out how to get back from the admin page to the edit page and the person on the other end of the line keeps telling me how the admin page works, he will only manage to escalate my frustration.

At the same time, the employee needs to navigate his own internal journey. New employees need to go through onboarding to understand the systems they'll use as part of exchanging service with customers. They need to be able to get help when they have problems with those systems and absorb changes to them. They need to be able to navigate both internal and external system outages.

Structural coupling between a self-steering organization and its customers happens through their respective journeys. In order for customer journeys to succeed, employee journeys must also succeed. Service quality depends not just on how well each journey works but also on how well they map to each other.

Value co-creation takes the form of conversations between internal and external journeys. When customers arrive at the point in their journey where they need help, the service must make it possible for them to ask for it. Asking for help triggers a corresponding transition in the support agent's journey. In order to successfully give help, support agents need to understand things such as:

- How to use the internal customer support system

- The application they're supporting

- The transitions that brought the customer to this point in his journey

- How to escalate problems they can't resolve on their own

Validating the quality of the customer's journey is merely the first level of QA in the coherency dimension. QA also needs to ask questions about employees' journeys, and about the mapping between employee and customer journeys. Ultimately, quality in the coherency dimension reflects how well internal and external journeys map to, or couple with, each other.

INTERNAL COHERENCY

In addition to the coupling between employee and customer journeys, service quality also reflects coupling between internal journeys. The various parts of the service organization need to converse with one another. Development needs to produce documentation and training that helps customer support agents understand the applications they're supporting. Support, in turn, needs to provide feedback that helps development understand customers' problems with the applications they're building.

DevOps represents a recognition of the need for internal journey-based structural coupling. Development's journey starts with understanding product requirements, moves through design, coding, and testing, and on to deployment. It ends with application monitoring, which drives new requirements thus turning the development journey into a cybernetic cycle.

Operations, for its part, also begins with product requirements, though expressed in operational rather than functional terms. From there, it moves through infrastructure planning and implementation, to application deployment, to its own level of monitoring. Just as with development, operational monitoring drives new requirements and restarts the operational journey.

Development and operations follow their own journeys, each of which needs QA for coherency like any other. Their journeys intersect at various points; traditionally, application deployment and incident escalation are the primary dev/ops intersections. One could think about DevOps as augmenting the coupling between the development and operations journeys. It adds new intersection points, including:

- Joint design reviews that address functional as well as operational concerns

- Multilevel monitoring that correlates infrastructure and application metrics

- Incident response that includes on-call developers

Just as QA in the coherency dimension validates the coupling between internal and external journeys, so too it needs to validate the coupling between development's and operations' journeys. In this way, QA fulfills a higher-order function. Beyond just validating how well a service works, it also validates how well the service provider's internal organization works. In particular, QA measures the success with which a service organization implements conversational practices such as Agile, DevOps, and LeanUX. In a classic cybernetic circularity, QA validates and thereby helps improve the quality of the conversational medium of which it's a part.

One could say that Agile, DevOps, and LeanUX all arose in response to quality problems with internal coherency. By adding new intersections between the dev/ops journeys, for example, DevOps improves their mutual coupling. From this perspective, identifying the need for these practices is part of QA, as is measuring the process by which an organization responds to the recognition of that need.

Of course, this process never ends. It is an integral, second-order component of continuous design. As part of validating their internal coherency, digital service organizations need to continuously test and adapt the components of their conversational medium. They need to use QA in the coherency dimension to steer their own steering mechanism. Taking this approach can help avoid cargo-cult or checkbox implementations. If the goal is to deploy features faster without compromising quality or to balance elegant UX with operational resilience, QA can ask questions like:

- Are we actually deploying features faster?
- How is change in speed correlating with change in quality?
- Are joint design/ops reviews actually impacting resilience?

Continuity

When I first signed up for my invoicing service, its feature set might only have covered 70% of my overall needs. Compared to my previous workflow, though, that 70% offered me compelling additional value. As a result, I was happy to work around the missing 30%. The convenience, performance, availability, and security I got from the service more than made up for the minor clumsiness of the required workarounds. It's this *compelling new value* that makes software service startups viable at all.

A year or two later, however, I'm not likely to remain satisfied with a service that addresses 70% of my needs. Now I expect it to expand to 80%, 90%, or even 100% of my needs. The workarounds that failed to dampen my enthusiasm when I first discovered the service now become significant annoyances. I initially judged the service by its ability to understand my core needs. Now I begin to judge it as much by its ability to understand and correct its own shortcomings.

At the same time as I expect the service to catch up with me, I don't stand still. My needs change and expand. Yesterday I wanted something to help me manage invoices. Today I want help managing expenses. Tomorrow I'll want help managing subcontractors. And so on, potentially without end.

Customers evaluate service quality in terms of features that are present or missing. They also evaluate it in terms of defects, aka bugs. Software organizations tend to treat bugs as things that exist at a given point in time. Customers, on the other hand, see bugs as indicators of the quality of a process. Is the software getting better over time, or worse? Is this release better than the last one? How long did it take to fix a particular bug?

Customers experience service as something that unfolds over time. This unfolding doesn't just reflect the customer's usage of the service. It also reflects the process by which the service changes in response to customer needs and competitive markets. The relationship between what a service is and how customers use it is as conversational as any individual acts of value co-creation.

Customers intuitively understand the cybernetic nature of service evolution. They know there's no such thing as bug-free software. They appreciate releases that fix bugs and add missing features. Bug-fix notifications are great brand-reinforcement moments. As much as customers want software that works, they also want a software provider who listens. Without knowing it, they judge service quality in terms of the provider's ability to self-steer in a manner compatible with their own self-steering. In other words, as the capabilities I need to successfully manage my business evolve, I expect my invoicing service provider's capabilities to evolve similarly as part of the successful running of its business.

Software-as-a-Service both enables and exacerbates the need to make continuous feedback an integral part of the customer relationship. Physical objects naturally throttle the rate of change. The materials used to make a car or a toaster are finite in availability. They must be procured from some place, then transported to the manufacturing plant, then physically transformed into a product. That product must then be shipped to a dealer. The customer must drive to the store, buy the product, then drive it home again.

Software as product shares similar traits. In this case, the friction created by change takes the form of operational requirements. When the new version of my invoicing software arrives on a CD in the mail, I have to install it. I need to make sure my systems will support it. I might need to upgrade the operating system or add memory or disk. I might even need to completely replace my systems with computers that have faster CPUs.

Software-as-a-Service largely removes this friction. It drives the cost of change for the customer toward zero. SaaS marketing touts operational benefits such as "we deploy upgrades so our customers automatically get new features." App stores for smartphones take the delivery of change to the next level. With iOS 7, Apple introduced auto-updating, which lets vendors automatically deploy new versions of their applications in a completely automated, transparent manner.

Thanks to this new, frictionless delivery model, customers have come to expect Software-as-a-Service to magically and continuously improve: to self-steer at the same rate as they do. Continuity is the dimension of quality that measures a service's ability to deliver change as an integral part of its normal operations. QA in the continuity dimension validates the quality and efficiency of the end-to-end process by which a service provider transforms feedback into new capabilities, whether those capabilities be new features, better performance, or improved documentation.

USER-CENTERED CONTINUITY

Software-as-a-Service drives the operational friction of change nearly to zero; it doesn't, however, remove cognitive friction. Continuous Delivery makes it possible to deliver change so quickly and frequently that it almost becomes truly continuous. People, on the other hand, can't absorb continuous change. They need help navigating the cognitive distance between one version of a system and the next.

A new user interface might provide a *better* or *easier* way to use a feature. In the context of change, however, the new user interface is *different*. The onus is on the service organization to incorporate the mental process of forgetting the old way and learning the new way as an explicit part of the customer journey. If it fails to account for this cognitive friction, the service organization risks having the customer equate *different* with *worse*.

Companies like Google have adopted a conscious approach to introducing change. They start by alerting customers to new features or interfaces *coming soon*. Then when they first introduce the change into production, they make its adoption optional. They give customers the opportunity to experiment with the new method and switch back and forth between new and old. Even after fully rolling out the new approach, Google gives users some level of control over when they fully accept the new interface and jettison the old one.

Continuity thus must concern itself with change absorption as well as change delivery. QA along the continuity dimension must test the quality of the absorption mechanisms that go along with any change. Is this change sufficiently complex or dramatically different to necessitate a gradual approach? If it does, how successful is the *absorption-ramp* functionality itself? Does it lower the cognitive load associated with a particular change, or just create more confusion? Does it succeed in making the change seem sufficiently desirable that the customer will invest the mental time and energy needed to adopt it?

INTERNAL CONTINUITY

Continuity impacts employees as well as customers. Software releases cause innumerable problems simply because they don't treat operational staff as release customers. Sysadmins can't diagnose and fix problems based on error log entries they don't recognize. Support agents can't help customers with features they've never heard of before.

The quality of internal delivery thus impacts the quality of external delivery. QA in the continuity dimension can't concern itself solely with the quality of the outbound delivery. It can't even concern itself just with the delivery touchpoint as part of the customer journey. In the post-industrial economy, delivering change is a critical part of value co-creation; QA thus must address the entire system behind it. It needs to validate the entire delivery system, including:

- The service provider's ability to deliver change
- Customers' ability to absorb it
- Employees' ability to absorb it
- The coupling between employee and customer delivery touchpoints

Continuous Design Quality

Customers judge failures in any of the dimensions of outcomes, access, coherency, or continuity as failures of the whole service. Service providers need to share this view of quality as inseparable across all four dimensions. They need to take QA beyond validating quality within each individual dimension. QA also needs to play a higher-level role. Its ultimate purpose is to illuminate the quality of continuous design as a whole.

The quality of a new feature, for example, depends on more than just how well it works. Its release can't compromise performance. It must at least fit within the existing customer journey; at best it will enhance that journey. It must be delivered in a way that can be absorbed by employees and customers alike. In addition, the organization must have ways to learn from it, and to optimize that feature and its internal processes by feeding those lessons back into the design process. In the complex, disruptive, post-industrial economy, the ability to self-steer through conversation is paramount. Companies need to be able to converse with customers, the market, and within themselves. From a cybernetic perspective, delivery is all-pervasive. Continuity doesn't just involve releasing new features. It can also mean changing infrastructure to improve performance. Game Days and *chaos monkeys* can generate changes in operational procedures. Customer feedback can even lead to changes in the delivery mechanism itself.

The principles of continuous design are intended to guide organizations in designing and operating high-quality services. The four dimensions of digital service serve as a mirror to those principles, allowing organizations to evaluate the effectiveness of their continuous design practice. QA serves as continuous design's internal feedback mechanism by posing questions such as:

- Does the service design for service, not just software, in a way that generates quality internal and external outcomes?

- Does the service minimize latency and maximize feedback in a way that increases continuity?

- Does the service design for failure and operate to learn in a way that maximizes access quality?

- Does the service treat operations as an input to design in a way that creates quality across all four dimensions?

- Does the service seek empathy in a way that generates effective self-steering?

SECOND-ORDER CONTINUOUS DESIGN QUALITY

Anything that changes the mechanism by which a company co-creates value with customers has continuity implications. Employee and customer journeys aren't static. People's needs and the environments in which those needs unfold continuously evolve. In order to adapt externally through a useful conversation with its customers, a service provider might need to adapt internally as well through useful conversations between the company's parts.

Internal conversations drive changes that must be delivered like any others. Consider the example of a company that decides to adopt DevOps. The adoption process is a cybernetic one. Employees need to understand, absorb, and participate in that process. The quality with which it unfolds will impact not just the company's internal efficiency but also its ability to serve its customers. People can tell when a corporate acquisition or internal procedural change has been clumsily handled. A complete QA practice will treat finding and exposing clumsiness as part of its job.

Post-industrial business is fundamentally about harnessing change. Post-industrial QA is about validating a service's ability to do so and helping them continuously improve. In order to succeed in that role, QA needs to address all dimensions of service at all levels of the self-steering conversation.

8

Building Quality In

In the industrial age, business processes could be tuned and then operated unchanged for significant periods of time. By contrast, the complexity and disruptiveness of the post-industrial economy requires continual adaptation. Businesses confront constant gaps in their coupling with the market and their customers. Service becomes a process of continual repair. From this perspective, *quality assurance* is just another name for the day-to-day functioning of the business.

Continuous design pushes this perspective even further. It challenges service organizations to acknowledge that design and development are never done, no matter how well tested their products are. Continuous design extends the QA activity of detecting gaps between actual and expected into production. In addition to making operations part of design, it makes QA part of operations.

Up to this point, we have naïvely used the term *quality assurance*, without examining its meaning relative to its use in the software industry over the past 40 years. Forward-thinking members of the QA industry have pointed out that "testing" can't actually assure quality. They have astutely recognized that what we call QA is merely the feedback half of the cybernetic loop. The job of the tester is not to assure quality, but rather to contribute to it by providing the information needed to guide accurate (re-)action. Productive self-steering requires useful conversations; testing validates an organization's ability to hear accurately as part of those conversations.

Some software quality practitioners have begun calling themselves *testers* instead of *QA engineers*. The word *testing* captures only part of what constitutes feedback in a digital service business. Probing, questioning, analyzing, and co-designing are equally important activities. For this reason, we will continue to use the term *QA* to refer to the spectrum of feedback mechanisms. We should remember, though, that the ultimate expression of quality assurance is nothing less than the continual process of business adaptation.

The four dimensions of digital service—outcomes, access, coherency, and continuity—describe various measures of quality. They present corresponding activities, which we've referred to as QA, that validate the achievement of quality based on those measures. These descriptions say nothing, however, about how or when these activities happen or who carries them out. The digital conversational medium reflects a particular strategy toward QA: its component practices all strive to intimately integrate cybernetic feedback rather than treat it as something separate and distant.

Optimizing Feedback

Imagine that you're rowing a canoe across a lake in rough weather. The wind is constantly changing speed and direction. As you move across the lake you encounter hidden currents of unpredictable strength and direction. Now imagine that the steersman in the back of the canoe only makes course-correction strokes every once in a while. The canoe will get pushed further and further off course. Your path toward your destination will be terribly inefficient, because you have to spend as much energy paddling your way back onto course as you do toward your destination.

This example illustrates the benefit of integrating testing within fast-cycling design, development, and operations processes. In addition to improving efficiency, though, tight feedback loops also improve adaptability and thereby competitiveness. The *winds* of disruption and the *hidden currents* of complexity don't just push businesses off course; they also move the destination. Every time a business looks up from their paddling, the dock at the other side of the lake has shifted position or changed size. The more quickly and continuously an organization can use feedback to detect changes in destination, the more energy it can devote to paddling in the right direction.

The transformation of IT in the post-industrial economy is changing how companies develop, operate, and plan software systems. New development, operations, and planning methodologies all take the approach of trying to build quality into the process of construction. In doing so, they deeply impact the role of QA.

Infusion Changes Design

In the digitally infused service economy, we conduct more and more of our daily activities through software interfaces. Companies like Nest turn physical control mechanisms such as thermostats into digital user interfaces. Services like HealthCare.gov make websites the primary mechanism for interaction between citizens and government.

As software becomes more central to our daily lives, user experience design becomes increasingly critical to service quality. Companies are embracing the value of user-centered design. Enterprise software vendors like IBM and SAP are touting their use of design thinking. Startups are recruiting designers as co-founders. Kleiner Perkins, a prestigious Silicon Valley venture capital firm, added the president of the Rhode Island School of Design as a partner (*http://bit.ly/1E7kTcI*).

The consumerization of IT is driving the need for good design into enterprises. Thanks to Twitter, Facebook, and other online consumer services, employees have become accustomed to high-quality digital experiences. Especially as they bring their own laptops, tablets, and smartphones with them to work, they begin to grow impatient with internal IT systems that are utilitarian at best and clumsy at worst. They expect corporate systems to reflect the same level of user-centered design excellence as the consumer software that shares their merged work/home devices.

Traditionally, we perceive designers as *controllers*. They figure out how to create solutions that will catalyze desirable responses in their users. A building should, for starters, not fall down. Beyond that basic requirement, it should satisfy its occupants both by keeping them warm and pleasing them aesthetically. A chair should be comfortable and elegant. A user interface should be *usable*.

These expectations all treat design as a process of creating things that act upon their users. This paradigm reached its climax with modernism, particularly in architecture, where designers saw themselves as building "machines for living."[1] User-centered design, by contrast, takes a deeply cybernetic approach to solving problems. Klaus Krippendorff, a renowned cyberneticist with a background in design, went so far as to define cybernetics, design, and *human-centeredness* in terms of one another. In his paper, "The Cybernetics of Design and the Design of Cybernetics" (*http://repository.upenn.edu/asc_papers/48/*), he

---

1 Le Corbusier, *Toward an Architecture* (Los Angeles: Getty Research Institute, 2007 [1923]).

called on designers to "conceive of their job not as designing particular products, but to design affordances for users to engage in the interfaces that are meaningful to them."

Krippendorff made the claim that good design can't happen without incorporating users' perspectives; in other words, without allowing itself to be controlled by the controlled. He saw the purpose of the design process as being not to build a thing but to engage in a conversation with a user. The resulting product would be an expression of that conversation.

In practical terms, user-centered design generates conversation through the practice of user testing. Incorporating testing within the design process provides feedback that validates the designer's beliefs about a solution's viability. Anyone who creates something is naturally biased toward believing in its usefulness. Furthermore, fully understanding what makes something meaningful to a user is extremely difficult.

User testing works by observing representative users while they interact with prototypes of a design in real-world scenarios. Controlled observation, along with post-session interviews, generates feedback that serves as input into refining a design. The design-test-refine loop might repeat multiple times over they course of a project. This iterative approach progressively narrows the gap between a solution's expected and actual usefulness.

User-centered design exemplifies the cybernetic strategy of "building quality in" by transforming construction into conversation. Integrating QA within an iterative feedback loop is key to this strategy. At its heart, it improves quality by increasing opportunities for meaningful empathy. It challenges designers to view their work as being as much about listening as about thinking, solving, or creating. It breaks down the dichotomy between traditional design questions such as "What do I want the user to do or feel?" and traditional QA questions such as "What does the user want me to do or understand?"

Service Changes Development

Delivering software as a service lowers the cost of change for customers by removing the onus of operations from them and putting it back onto the vendor. As customers become accustomed to the service model, their patience with slow rates of change shrinks. Eventually they come to expect nearly instantaneous response from the service provider. They begin to judge software services by how smoothly and continuously they can repair gaps between actual and expected quality, whether that means fixing bugs or adding new functionality.

This escalation of customer expectations forces software service providers to adjust the way they produce software. Long release cycles are no longer feasible. Continuous repair requires the ability to deliver change more often in smaller quantities. The need for Continuous Delivery creates a conundrum. A high-velocity process is more sensitive to noise. Trying to improve quality by going faster can, by its very nature, degrade quality. We've all had the experience of making mistakes because we're in a rush.

In order to increase speed without sacrificing quality, a software delivery methodology must find ways to engineer noise out. Handoffs and translations are prime sources of noise in the software lifecycle. The waterfall model maximizes both. Handoffs take place between marketing, development, and QA. These handoffs involve translations between marketing requirements documents, design specs, and test plans. Validating the accuracy of these translations becomes a QA activity in its own right.

Agile is a response to the need for more continuous software repair. Its central strategy is to reduce the size and increase the frequency of the design-develop-test lifecycle. Over time, Agile has discovered and incorporated several powerful techniques to remove noise from the iterative lifecycle. These techniques all involve breaking down boundaries between development and QA people, practices, and tools.

EMBEDDED QA

Traditional software organizations stress the importance of separating QA from development. They view QA's role as being that of an external auditor. They believe in an adversarial relationship between development and QA. This structure reflects the assumption that development naturally produces poor-quality code. It also reflects the belief that doing so is acceptable. Rather than trying to improve quality within development, these organizations rely on QA to correct development's poor work from the outside.

Agile turns this perspective on its head. It puts the onus on development to produce not just quantity but quality. It does so by embedding QA within development. Regardless of the formal organizational structure, Agile creates opportunities for QA engineers to work intimately with developers on a daily basis. They sit together, attend meetings together, participate in planning sessions, and critique software designs.

At Apple, I participated in a transition from an adversarial model of the QA-development relationship to one based on intimate partnership. QA engineers were relocated from a separate floor to sit within the development team pods instead. They started reading and finding bugs directly in the source code. Far from resenting the intrusion, developers began to actively seek out code reviews from the QA engineers on their teams. Quality skyrocketed, as did QA engineers' self-esteem, which only served to increase their value.

TEST-DRIVEN DEVELOPMENT

The most efficient way to produce good code is to detect bugs during the development process. The most efficient way to maintain code quality is to detect bugs while making changes. Unit testing is a technique for accomplishing both of these aims.

Unit testing works by making test construction part of development itself. Instead of concentrating solely on writing code and leaving the writing of tests for QA engineers, developers also write tests for the code they're building. This tactic has multiple benefits. It makes developers accountable for the quality of their own code. It helps them find bugs immediately before they submit their code for team-level integration and testing activities.

Unit testing also creates a dynamic framework for regression testing. Maintaining unit test suites alongside the code and adding new unit tests as code is added and changed makes it possible to detect breaking code changes. New code should not break old unit tests. If it does, developers know where to look for integration problems.

Unit testing helps uncover logic bugs. By itself, though, it doesn't help uncover design bugs. To properly answer the question "Does the code work as expected?" one needs to understand what "as expected" means. *Test-driven development* (TDD) is an often misunderstood technique for addressing that question. TDD involves writing unit and integration tests before writing the code. The standard approach is to write a test that fails simply because there's no code to implement it and then write the missing code.

The value in TDD lies less in ensuring tests are being written than in forcing developers to think about their code from the outside in instead of the inside out. By starting with the test, they start by thinking in terms of how their code will be used. What purpose will it need to serve? Only after formally expressing the code's purpose through the test does the developer express, through the code itself, how that purpose will be solved. Writing tests first helps developers "write the right code, not just write the code right."

EXECUTABLE REQUIREMENTS

The concept of *executable requirements* is a method for collapsing the distance between requirements, specs, and tests. It reflects the insight that tests are a very compact representation of requirements. If you can express requirements in testable language, you can minimize the need for translation between users, marketing, development, and QA.

Behavior-driven development (BDD) is an extension of TDD that moves it into the realm of executable requirements. BDD uses simple, human-readable language for expressing requirements as tests. By describing desired changes in terms everyone can understand, BDD fosters useful conversations between all the stakeholders in the development process.

The key to gaining benefit from BDD is not to treat it just as a more compact way to write down a spec. Instead, teams should use it as a tool for collaboratively developing scenarios. When users, product managers, developers, and testers flesh out scenarios together, they can hear and learn from one another's feedback. Developers can hear users say things like "That's not what I meant," or "That's not how I want it to work." Conversely, users can respond to developers asking questions like, "What if we did it this way?"

Requirement bugs are the best kind. Uncovering an inaccurate requirement, or exposing a misunderstanding about a requirement, saves the entire team from wasting time designing, building, testing, and releasing the wrong thing. When approached as a conversational tool, BDD becomes a form of QA by helping to detect understanding-level bugs.

CONTINUOUS INTEGRATION

The best-designed, best-written tests do no good if they're not executed. Ideally, tests get run against every code change. Even with armies of QA engineers, manual test execution has little hope of keeping up with high-velocity development lifecycles. Not only do important tests get omitted, but they might get improperly executed due to human error. False positives ("all executed tests pass") create the worst possible situation for an organization trying to deliver quality code fast.

To provide genuinely useful feedback, tests need to automatically run whenever needed. Continuous Integration (CI) automates the process of detecting changes and building code, running tests, and reporting results against those changes. Testing every change uncovers defects as quickly as possible. It also simplifies the repair process by isolating defect sources. It's much easier to diagnose the cause of a bug if you can associate it with a single, small code change.

In order for Continuous Integration to be truly effective, Agile teams need to take it seriously as a cybernetic mechanism. Code-writing and test-writing need to happen as synchronously as possible. Writing unit tests after the fact severely compromises their usefulness. The same holds true for higher-level integration tests. A situation where QA engineers write integration tests the iteration after developers write code is an unfortunate and all-too-common Agile team dysfunction. A more effective model pairs developers and testers so that they write code and tests and review each other's work prior to committing any of it.

Even more important is the need to honor the feedback that a CI system provides. CI contributes to high-velocity quality by uncovering defects early and often. Failing to address defects immediately upon their exposure loses the value of doing it at all. A team that leaves a build broken for days on end might as well fall back into the old model of big-bang testing.

Just as user testing serves no purpose if designers don't adjust their work in response to it, so too CI does no good just because a development team automatically runs tests every day. Useful conversation requires not just listening but also responding to what's been heard as an integral part of the production process. To be truly Agile, any particular Agile methodology needs to honor this dictum.

Complexity Changes Operations

The post-industrial business model breaks down boundaries between IT systems, and between IT and the rest of the organization it serves. As systems, processes, and organizations become more interconnected, their structure changes from complicatedness to complexity. Complex systems require a dramatically different strategy for dealing with failure. Attempts to eradicate it are futile and counterproductive.

Complex-systems management needs to strive for resilience rather than stability. On one level, resilience means being able to operate without being perturbed by component failures. On another level, it implies the ability to learn from them.

QA typically functions as a mechanism for detecting and removing failures. Development organizations pursue the mirage of defect-free software. Common to all software QA methodologies is the strategy of maximizing failure in test environments in order to minimize it in production environments. To some degree, this strategy still holds. It is possible to build robust software components. It isn't feasible, however, to fully test or even understand the interactions between components or between the complex systems that emerge from them.

Post-industrial IT thus needs to incorporate resilience into its definition of quality and thus into its QA practices. As QA moves from validating software to validating service, it needs to test the ability to survive failure as well as the ability to learn from it. Particularly in the latter case, QA must transform its relationship to operations in a profound and counterintuitive way.

VALIDATING DESIGN-FOR-FAIL

Stress-testing is a common software QA technique. It involves seeing how a software product behaves under unexpected conditions. If, for example, an application requires at least a megabyte of memory to run, a stress test would try to run the application with only half a megabyte.

Testing design-for-fail extends stress testing into the service domain. Design-for-fail implies a service's ability to survive, sometimes through graceful degradation, in the face of component failures. How does the service behave if the database is down? If the internal ERP system is down? If the external mapping service is down? If any of these systems is operating at an impaired level?

Design-for-fail validation needs to concern itself with resilience on multiple levels:

- How well do application components survive infrastructure failures?

- How well does the user experience survive application component failures?

- How well does the service organization survive system failures?

Validating design-for-fail requires new test analysis skills. A good test plan reflects a deep understanding of the system under test, its components, and their relationships. Complexity forces QA to enter the realm of systems thinking. In order to exercise potential failure modes within complex systems, QA must develop the ability to think in terms of relationships, both between components and between hierarchical system levels.

VALIDATING OPERATE-FOR-FAIL

Software services are not purely technical systems. Regardless of their level of sophistication or automation, they still require human operators. The unfeasibility of perfectly masking component failures necessitates the capability for human intervention. Service design needs to incorporate mechanisms by which operators can detect and respond to failure.

System operators need to be able to respond to failure on two levels. Design-for-fail might mask some failures from customers; operators still need to heal those failures in order to bring a system back into an optimal state. In a complex system, though, design-for-fail can never be perfect. There will always be failures that evade its capabilities. Outage response is thus a critical capability of any IT organization.

Game days (as described in Chapter 4) bring QA into the domain of validating an organization's ability to operate-for-fail. Done properly, they test the ability to respond to maskable as well as unmaskable failures. In order to design good game day exercises, QA needs to stretch its systems-thinking skills even further. In particular, a game day needs to test more than just IT's ability to detect and repair failures. It also needs to test the processes for communicating failure detection and response, both within the service provider's organization and with customers.

VALIDATING LEARN-FROM-FAIL

In order to validate a service's design-for-fail and operate-for-fail capabilities, QA needs to extend its understanding of what constitutes a system. Tackling the ability to learn from failure, on the other hand, requires completely changing your basic understanding of QA. The learn-for-fail mindset doesn't just posit that failure is unavoidable. It also proceeds from the insight that complex systems can't be truly simulated. The only place one can accurately model a complex software system is in its actual operating environment. There is no choice but to test in production.

This conclusion flies in the face of everything software QA has learned and holds dear. Testing in production is normally treated as evidence of a breakdown in quality. Letting defects make it through the production is bad enough. The last thing one wants to do is intentionally provoke them by running tests in customer-visible environments.

Learn-for-fail fully collapses the distance between operations and QA. The Netflix chaos monkey functions as a testing tool, probing and stressing systems to expose faults. At the same time, it serves as part of the company's operations toolset, just like monitoring, load balancing, or database replication.

The goal of learn-for-fail strategies is, quite simply, to learn from failure. From a QA perspective, the chaos monkey only works if it does in fact generate failures in production environments. Learn-for-fail validation involves testing:

- The ability to expose useful failures in production
- The organization's ability to recognize and respond to them
- The extent to which resilience improves as a result

VALIDATING DISTRIBUTED ARCHITECTURES

Complex systems don't respond well to centralized, top-down control strategies. Decentralized architectures such as microservices improve scalability and resilience by distributing control. In the process, however, they ironically increase QA complexity.

The decentralized nature of a distributed architecture helps scalability and resilience by dispensing with the need for global knowledge or coordination. That very characteristic, however, makes comprehensive testing infeasible. In order to reap the benefits of distributed control, the organizational structure needs to mirror the architectural structure. Each component of the architecture must view itself as an autonomous service. As an autonomous service, it must incorporate development, operations, and QA perspectives within itself. It must address both its internal functioning and its interdependencies with other autonomous services from all three perspectives.

To some degree, a decentralized architecture pushes design-for-fail to its ultimate conclusion. Design-for-fail maximizes autonomy. Validating the extent to which a component succeeds at autonomy is a key QA function. At the same time, though, each service needs to support its dependents from a QA perspective as well as an operational one.

This requirement impacts the continuity dimension in particular. Services that depend on me are my customers. As part of testing changes to their own systems, they need to test their services' interaction with mine. I therefore need to manage my own continuity not just for the benefit of my operational customers but also for the benefit of other services' deliverability as well.

I can satisfy this need by providing a persistent integration environment or through on-demand simulation environments that dependent services can deploy at will. The specific implementation is less important than the relationship between service organizations. The key point is to realize that my QA process, while autonomous, must converse with other autonomous QA processes. The quality of the structural coupling between autonomous parts of an organization applies to QA as much as to any other aspect of digital service.

Adaptation Changes Planning

Just as complexity changes the relationship between QA and operations, so too does adaptive service change QA's relationship with planning and design. The product delivery model of software forces companies to finish planning, designing, building, and testing a feature set prior to releasing it to customers. Software-as-a-Service, on the other hand, makes it possible to treat the release process as an opportunity for continual experimentation.

The industrial model puts planning at the front of the product development lifecycle. Market research and design precede product implementation and sales. By driving down the cost of change for service providers, Continuous Delivery makes it possible to transform the planning lifecycle into a cybernetic loop that incorporates development and operations.

Feature release techniques such as A/B testing, canary releases, and demographically targeted releases transform product marketing. Traditionally, product organizations answered the question "Which version should we release to the market?" in advance. Thanks to Agile delivery practices, they can turn that question into an ongoing conversation with real customers. They can, in effect, conduct marketing-level testing in production just as operations conducts technical-level testing.

Continuous Design Changes Everything

The ability to test in production steps onto the slippery slope to continuous design. For the most part, the strategy of building quality into the design, development, and operations of digital systems still inherits their physical-world ancestors' emphasis on iterating toward a finished product. A cybernetic marketing loop, though, necessarily incorporates design, development, and operations. It turns design into continual experimentation. Perpetual development, coupled with the ability to test in production, challenges the idea that a design can or need ever be *done*.

Testing in production, whether in the form of A/B testing, chaos monkeys, or simply application and system monitoring, provides information that can feed back through the organization to create a circular design process. In order to contribute to continuous design quality, however, that feedback has to happen in a meaningful way. Minimizing latency without maximizing feedback doesn't help companies co-create value through digital conversations. The value of twenty-first-century IT techniques such as Agile, DevOps, and Continuous Delivery doesn't come from letting companies push change at customers more rapidly or continuously. It comes from helping them steer more effectively by reducing the delay in receiving information *back* from those customers.

The importance of a holistic approach to service is a central part of this book's premise. Quality measures the four dimensions of outcomes, access, coherency, and continuity inseparably. Useful conversations between a company and its customers depend on useful conversations between the company's parts. Feedback thus must happen on the right levels and must flow to the right places.

As discussed in Chapter 3, the digital conversational medium enables multi-level operational feedback, including infrastructure, application, and user-behavior monitoring in addition to A/B testing, customer support, and social media. In combination with operational learning practices such as game days and chaos monkeys, along with Continuous Delivery metrics, these metrics provide visibility across all four dimensions of service.

None of these feedback sources have value, though, unless they are treated as first-class inputs to design. True service quality depends on the ability to make correlations across service dimensions and between organizational components. A user-interface change that pleases users while seriously degrading database performance needs further design. It represents a disconnect between frontend and backend development. Validating the organization's ability to fully absorb and respond to feedback is thus QA's most subtle role, and the source of its greatest value.

Pervasive QA

Software quality improvement methodologies often incorporate the idea of "shifting QA to the left." According to this concept, the earlier in the lifecycle you test, the easier, faster, and cheaper it is to fix defects. Involving QA in the requirements process is the ultimate expression of shift-left testing. Requirements bugs are the cheapest to fix because finding them avoids writing the wrong code in the first place.

Agile development techniques such as unit testing and Continuous Integration are examples of shifting QA to the left. Complexity, on the other hand, forces us to shift it in the opposite direction. No longer can we assume that we can finish testing before we release changes to customers. On the level of operations, we shift QA to the right because we have no choice. On the level of marketing and design, we shift it to the right because we can.

In both cases, shifting QA to the right lets digital service organizations change their focus from *construction and operation* to *continuous learning*. The truth is that post-industrial companies need to shift QA in both directions simultaneously. In order for IT to function as a digital conversational medium, it needs QA to become pervasive throughout the continuous design lifecycle.

When we say that a cybernetic process "collapses the distance" between action and feedback, we mean that it defines a larger process that incorporates them both. In order for any cybernetic mechanism to operate optimally, both halves of the pair must work efficiently. The more quickly, continuously, and smoothly you can adjust the aim of your gun, the better your chance of shooting down a fast-moving enemy plane. Adjusting your aim requires both detecting the error in the turret's current position and correcting its position based on feedback. A wildly oscillating turret will compromise the ability to shoot down the plane just as much as poor information about its flight path will.

Shifting QA to the right maximizes access to feedback. It provides the most realistic feedback by gathering it from real customers in real situations. It provides the greatest quantity of feedback by letting go of the notion that one must stop testing upon release to production.

Shifting QA to the left conversely maximizes the ability to respond to feedback. It speeds up the change delivery process by addressing defects in the quickest and cheapest possible way. By the same token, it increases the accuracy of what's being delivered. Shifting to the left thus minimizes the oscillation of the gun as it tries to repoint itself.

IT's purpose as a conversational medium is to allow post-industrial businesses to conduct useful conversations with customers by continually changing existing situations into preferred ones. Achieving that purpose relies on building the detection of error into the continuous design process itself. No longer can QA function as a segregated, adversarial component of the development process. It can't even view itself separately from operations. Instead, QA must become pervasive throughout the entire digital service organization. A continuous design practice that drives truly effective self-steering will be one that builds quality into the very fabric of the conversational medium.

9

From Quality
Assurance to Quality
Advocacy

In a traditional software organization, responsibility for quality assurance activities falls to a dedicated QA team. The QA team is intentionally separated from design, development, and operations, often reporting up through an entirely different branch of the organization. QA engineers focus entirely on building, running, and reporting on tests. They interact at arms length with marketing and development through the mechanisms of the specification, build, and defect-tracking system.

QA traditionally plays an adversarial role within the software delivery lifecycle. It is up to them to "certify" a release's quality. They often have, or are perceived to have, the authority to stop a release from being delivered to customers.

Pervasive QA dissolves the boundaries that separate QA from marketing, design, development, and operations. It disperses responsibility for many QA activities across other teams. Marketers design A/B tests and interpret the results. Designers conduct user testing sessions. Developers write unit tests. Operations runs game day exercises, reads monitoring dashboards, and deploys chaos monkeys.

Automation further erodes dedicated QA. Software QA teams spend a surprising amount of their time on tedious, repetitive activities. They manage test suites in spreadsheets and execute them by hand from the command line. They manually correlate test results and hand-generate reports after each test run.

Automating the test execution process is central to the value of Continuous Integration. CI servers ensure that every test gets run on every build. They automatically invoke test result report generators and distribute their output to every-one on the team. As one of its many benefits, CI can reduce QA engineers' manual workload by up to 50%.

The Erosion of Traditional QA

The move toward pervasive, highly automated QA might give the impression that post-industrial IT organizations can dispense with QA teams altogether. Doesn't the very notion of a dedicated QA engineer run counter to the cybernetic princi-ple of integrating action and feedback? When quality is built in, it seems, you shouldn't need anything extraneous on the outside.

This notion is a victim of the myth of the full-stack engineer. The Agile emphasis on self-organizing teams, if not properly understood, contributes to this myth. A rigid interpretation of self-organization posits that every team mem-ber should be able to play every needed role. Otherwise, how could the team organize itself flexibly?

According to this view, every team member must be equally qualified to complete every task needed to deliver customer value. They have to be able to understand user needs, design and write code, develop tests, manage databases, architect and operate infrastructure, and so on. Some Agile purists go so far as to claim that, if the word *tester* is in the job title of anyone on your team, then you're "not doing Agile."

Taking such an extreme view sets teams up for failure. If you extend the con-cept to digital service, the entire team also has to be skilled at service design, user documentation, and customer support. The likelihood of finding any one person, let alone an entire team of people, who have equally strong expertise in such a wide range of skills is remote at best.

T-SHAPED PEOPLE

If it doesn't work to have a team full of people who can do everything, what is the alternative? Continuous design and the pervasive QA it needs require intimate relationships between practitioners across disciplines. Organizational and proce-dural silos generate waste, slowing down the cybernetic loop and introducing noise that impairs its effectiveness. What's needed is an approach that combines deep expertise with strong interconnection. The alternative to teams of mythical full-stack engineers is teams of *T-shaped people.*

The phrase *T-shaped person* was coined by Tim Brown, CEO of the design firm IDEO and author of the book *Change By Design* ([brown2009]), which has played a key role in popularizing design thinking. In an interview with *Chief Executive* (*http://bit.ly/1E7pcFo*), Brown defines a T-shaped person as having two concurrent, complementary qualities. These two qualities respectively correspond to the vertical and horizontal strokes of the letter T:

- Depth of skill in a specific area
- Disposition for collaboration across disciplines

Brown goes on to define the horizontal stroke as depending on empathy. A T-shaped person has an enthusiasm and appreciation for other disciplines and a desire to engage with them. This desire is driven by the T-shaped person's ability to see things from other perspectives.

T-shaped people are the ideal team members in post-industrial service organizations. Self-steering relies on empathic conversations across disciplines. Empathy requires the ability to reach out to something different and other than oneself. From this perspective, the mythical full-stack engineer is not just impractical but also unnecessary, and even undesirable.

THE TESTER'S PERSPECTIVE

In order to deliver quality digital service, post-industrial companies need a broad range of skills. They need people with these skills to work closely with one another. Each skillset represents a complementary perspective. To a large degree, quality depends on how well these perspectives come together. Excellent service consists of well-targeted features expressed through excellent design, implementation, business and technical operations, and support. Continuous design binds these disciplines and activities together into a circular loop; it doesn't mush them into an amorphous blob.

Skills and perspectives need to come together across design and engineering to create a successful conversational medium. The cybernetic nature of this medium means that designing, generating, and interpreting feedback need to be integral parts of all of its activities. At the same time, however, the ability to think deeply about testing is a perspective unto itself.

Most disciplines train their practitioners to think in terms of action. Marketers plan. Designers design. Developers code. Operations deploys. QA is the only discipline that focuses primarily on feedback. QA engineers sometimes get a bad rap for being negative, pessimistic, and destructive. Whether deserved or not, this reputation reflects the fact that most disciplines think in terms of expected results. Designers think about how they want a web page to look. Developers think about how they want it to work. Marketers think about what they want it to make a customer do. QA, on the other hand, is the discipline that concerns itself primarily with the gap between expected and actual behavior.

CONFIRMATION BIAS

QA can play an especially important role as an antidote to confirmation bias. No matter how empathic we are, no matter how much we test our designs, we all are instinctively attached to our own creations. Designers want to believe their websites are beautiful. Developers want to believe their code is fast and bug-free. System administrators want to believe their infrastructures are robust.

QA, on the other hand, is attached to the customer. This characterization might sound glib, but we can easily see it in action. When HealthCare.gov melted down, the first thing people said was, "It wasn't tested properly." When bugs make it into the field, QA gets blamed. QA cares about customer satisfaction not just out of altruism but as a matter of self-preservation.

User-centeredness means that the customer's experience wins out over the designer's. QA functions as a representative for this perspective. Its bias toward the customer provides exactly the feedback mechanism needed to achieve real user-centeredness.

QA's Changing Role

Pervasive QA need not lead to the conclusion that QA as an explicit discipline is unnecessary. Instead, by building appreciation for quality into the organization as a whole and by removing the need for tedious, repetitive manual tasks, it can free QA to play a deeper, more valuable role. This new role embeds QA engineers as participants within cross-functional teams of T-shaped people.

What are the characteristics of the vertical stroke that defines the T-shaped QA engineer? In a sense, QA is the mirror image of the other roles that make up the conversational medium. Where other team members specialize in various types of design and construction, QA specializes in designing and constructing feedback mechanisms. Whereas other team members instinctively look out from within, QA's customer attachment helps them look in from without.

As members of T-shaped teams, QA engineers need to express themselves via the horizontal as well as the vertical stroke. They can't just sit in the corner and tell everyone else they're wrong. They have to engage positively and empathically with the other roles within the team. Taking an interest in the things they're testing and the people designing and building those things helps QA contribute to making quality more pervasive.

T-shaped QA engineers can participate in design and architecture sessions, code reviews, game days, and user testing. They can help other team members think about testability throughout the continuous design lifecycle. They can infect the entire lifecycle with their caring for the customer. By integrating themselves within the teams that build things rather than sitting apart in judgment, QA can change other disciplines' perception of it. In particular, they can change design and development's view of them from suspicion to trust. As a result, other disciplines begin to seek out QA's input instead of avoiding it as an annoyance.

When QA engineers become trusted resources, they multiply the power of their role within digital service organizations. No longer is their value limited by the number of tests they can write and run. As integral members of continuous design teams, their role transitions from one of mechanical test execution to one of representing the principles of continuous design from a quality perspective. In particular, QA can function as an trusted advocate for three key quality drivers: service, not just software; built-in quality; and resilience and adaptation.

SERVICE NOT SOFTWARE

Customers judge digital service quality inseparably across the four dimensions of functionality, operability, coherency, and deliverability. All too often, though, delivery organizations forget this fact. Software's product legacy creates a strong temptation toward a tunnel-visioned focus on functionality.

We are still entering the age of the service economy and learning what service really means. Thinking in terms of customer journeys, jobs-to-be-done, and value co-creation is still foreign to us all. Just as we had to learn to value user-friendliness as a guiding principle for user interface design, so too we have to learn to align ourselves with customer outcomes over our own constructions.

Reminding delivery teams of the need to build, operate, and support service, not just software, is a key part of QA's new role. This reminder applies to all aspects of the delivery lifecycle. In the context of requirements, we need someone to ask, "What customer job-to-be-done does this feature address?" In the context of design, we need someone to ask, "How will the customer learn and onboard this feature?" In the context of deployment (the intersection of development, operations, and support), we need someone to ask: "How will the unexpected appearance of this feature impact the customer? Will they need the ability to control their adoption of it? If so, how?"

Service over software impacts more than just the interface between the service provider and the customer. It also necessitates holistic internal alignment. Coherent customer journeys depend on coherent business processes. Customers don't arrange their journeys based on corporate organizational structures. In order to meet the customer where they are, a company must ensure that its internal relationships accurately and efficiently map to its external relationships.

The emergence of DevOps reflects this reality. Writing good code is necessary but not sufficient for a software service company. The code must be operable and supportable. From the perspective of the customer journey, users don't just need to use a feature; they also need the ability to get help when they're having trouble with it. In order for a customer to get help, a support agent or technical documentation writer has to be able to provide help. Support and documentation are development's customers just as much as end users.

On one level, validating internal alignment means addressing operations and support needs as part of the software test plan. On a deeper level, though, it means validating the nature of the relationship between teams. The most efficient way for development to address support's needs is to engage them as part of the requirements process. In addition to telling development what is necessary in order to support a given feature, support agents can call on their experience with customers to provide feedback on the feature itself.

By facilitating empathic conversations between development and support, QA can make both better. QA's new role includes ensuring both that these conversations happen, and that they are in fact empathic. Empathy is the oil that lubricates the flow of understanding between teams. Mutual, dynamic understanding allows the components of a complex service to continuously organize themselves around the unfolding customer journey.

When marketing, design, development, operations, and support all share the same understanding of a new customer need, they can efficiently and accurately collaborate to fulfill it. Unfettered, empathic flow across disciplines is what characterizes a successful conversational medium. Helping cross-functional teams see and remove obstacles to that flow is an important new QA function.

BUILT-IN QUALITY

Pervasive QA relies on making the QA mentality pervasive throughout the service delivery lifecycle. Someone needs to ensure that the infusion of quality into implementation truly happens on an ongoing basis. The simple fact of writing unit tests, running Continuous Integration builds, and releasing A/B versions of features doesn't guarantee that any of those activities will be useful. We must always guard against the temptation for cargo-culting.

Someone needs to question the quality of each of the specific built-in quality activities and artifacts. Do developers write useful unit tests or just make sure that every code module has one? Does the development team take Continuous Integration seriously and stop development to fix the build whenever it's broken, or do they start to ignore build failures and let them pile up for later attention? Do operational dashboards radiate meaningful information about the health of the service, or do they just create pretty flashing lights on a monitor on the wall? Do designers and marketers use A/B tests to measure results against truly useful parameters or just vanity metrics?

QA can fulfill this need by serving as a chaperone for building quality in. Its primary focus on feedback lets it view test activities and artifacts on their own merits. Developers instinctively treat code as their primary purpose and tests as secondary. Designers and operators do the same in their domains. QA engineers look at things from the opposite vantage point. They can use their training in test design to evaluate the quality component of the other disciplines' work.

Just as developers review one another's code, QA engineers can review unit tests. They can include metrics such as how long builds stay broken as part of their quality reports. They can help operations understand which monitoring reports are useful to whom and for what reasons. They can review A/B testing metrics the same way they would test plans.

Building quality in doesn't mean making defects disappear. The supplanting of waterfall by Agile demonstrates the futility of that hope. Moving quality forward means moving testing forward. The goal is not to pursue a mythical holy grail where there are no bugs but rather to make it possible to find bugs as soon as possible at the highest possible level of abstraction. The less energy that's been expended toward creating something that contains a defect, the easier and cheaper it is to remove that defect.

In order to achieve that goal, one must value testability along with ease of creation. QA can serve as an advocate for testability. At all levels of the delivery process, QA engineers need to ask, "How will we test this hypothesis?" The hypothesis in question can be everything from the basic premise behind a proposed new service, to the belief about what job-to-be-done it serves, to a feature spec, to a rollout plan, to an operations and support plan.

RESILIENCE AND ADAPTATION

Quality measures an organization's ability to control its destiny by creating things that remain fit for purpose over time. In the complex world of post-industrial business, fitness for purpose must include resilience and adaptability. These characteristics apply at all levels of digital service:

Infrastructure
Can the underlying IT platform respond to component failures and changes in load?

Application
Can the application respond to infrastructure failures as well as component failures and changes in load at its own level?

User experience
Can the user interface degrade gracefully in the face of application failures?

Service support
Can human operations respond to failures in the digital systems they support?

The need for resilience and adaptation even applies at the level of the service delivery organization itself. In a microservices architecture, what does one service team do when another team upon which it depends doesn't meet its deadlines or doesn't produce high-quality code? If the company detects signals that the market is changing, how does it reconfigure itself to meet new customer needs? Issues of component failure and changing load impact organizational structures and procedures just as much as they do software and hardware.

QA can help service delivery organizations achieve resilience and adaptability on two levels. QA engineers can help teams design and execute specific resilience tests. They can serve as advocates for resilience analysis throughout the delivery lifecycle. During technical design sessions, they can ask questions such as "How will this proposed design change impact our existing failover design?"

QA engineers can provide reminders about the need to address resilience at each level. They can ask UX designers, "What happens to this page if the service it relies on isn't available?" They can ask service designers, "What happens if the reservation system is offline?"

On a deeper level, though, they can serve as the advocate for resilience over stability in the face of complexity. Whereas stability-based strategies attempt to engineer out failure, resilience accepts the inevitability of failure. It further accepts the impossibility of perfect design or even understanding. To some degree, QA asks questions in order to generate answers. Beyond that, though, it asks questions to provoke humility. QA can provide tremendous new value by continually reminding the entire service delivery organization that everything it does is an experiment.

Success in a complex system consists of conducting useful conversations with others in your environment. Without recognizing that your actions are inherently incomplete and imperfect, you cannot hope to hear feedback from others in your environment. Without accurately hearing feedback, you cannot hope to take useful next actions. Instead, you're doomed to wander and wallow. As much as anything, QA's new role consists of helping digital service-delivery organizations make listening a first-class activity.

Continuous Quality

The waterfall model of development made QA into a tail wagging the dog. Testing happened at the end of the lifecycle. It functioned as a bottleneck that slowed down the release process. Testers saw their job as putting their fingers in the dike of buggy software leaking into production.

Product managers and developers alike viewed testers as obstacles to delivering customer value. As a result, testers felt hated by everyone. They were prized for their negativity, and lived up to their reputation as a defense mechanism.

The truth is that the waterfall model tasked QA with an impossible job. The infeasibility of this job can be seen in the very name *quality assurance*. In their role as detectors of error, testers can't assure quality. They can only provide the feedback needed to make course corrections. The willingness and ability to listen to feedback, and openly respond to it, lies with the entire organization.

QA's new role transforms its job from *quality assurance* to *quality advocacy*. As T-shaped members of cross-functional teams, they focus on helping ensure a continuous focus on quality throughout the entire software lifecycle, and beyond to the entire service operations lifecycle. Now, instead of serving as judges set apart from creation, they become integral aids to it.

Pervasive, embedded QA helps the digital conversational medium achieve continuous quality on multiple levels. Testing happens continuously within each discipline, and during every phase of service design, development, and operations. No longer is testing cordoned off from either the left or the right extremes of the lifecycle.

Continuous testing enables continuous listening. User testing, unit and regression tests, resilience testing, production monitoring, game days, chaos monkeys, and blameless postmortems all generate information that can be used for self-steering. Integrating the generation of this information into design, development, and operations creates a truly circular environment. Tools such as *information radiators* bathe service delivery organizations in feedback. It transforms the entire organization's perspective from one of creation and delivery to one of conversation.

Continuous listening enables continuous repair. Gaps between *actual* and *desired* make themselves known quickly, and close to those most able to repair those gaps. The more immediate the feedback, the smaller the gap being identified and thus the easier it is to repair. Service delivery becomes a smoothly rolling wheel, and self-steering becomes fluid and responsive.

In addition to being continuous through time, quality also must be continuous across the four dimensions of digital service. Pervasive QA, expressed through T-shaped teams, provides global visibility into feedback across all four dimensions. It empowers organizations to achieve continuous quality on all levels of the digital brand conversation.

CONTINUOUS QUALITY AND CONTINUOUS DESIGN

To a certain degree, continuous design fosters continuous quality by its very nature. Through its circular process, it repeatedly tests and refines service quality. The focus on addressing the gap between current and preferred as the essence of service builds in an appreciation of the importance of feedback.

On another level, though, continuous design depends on continuous quality. Pervasive QA is the aspect of continuous design that infuses this appreciation of feedback into the entire organization. It is QA that maintains the view that values other over self, and listening as an equal partner with acting.

When pervasive QA infuses continuous design with continuous quality, it makes testing and repair an integral part of creation. Effective continuous design requires that depth of infusion. Just as post-industrial business relies on frictionless value delivery, so too it relies on frictionless feedback. Otherwise, the company starts to fall behind the customer. The fidelity of the information it uses to respond to the market begins to grow stale and degrade.

SECOND-ORDER CONTINUOUS QUALITY

A company's ability to make, and keep, the right promises is the ultimate determinant of service quality. In the post-industrial world of service, infusion, complexity, and disruption, promise-keeping and promise-making both must be continuous. Self-steering requires the ability to adapt not just what you do but also how you do it. Components of autopoietic organizations must cybernetically manage the quality both of the services they provide to one another and of the ways they relate to one another.

Continuous empathy is the ultimate driver of value co-creation. It focuses an organization's emphasis on customers' needs over its own products and processes. From this perspective, the organization treats its internal mechanisms as useful only to the extent that they meet others' needs.

QA's most subtle role consists of helping organizations maintain an empathic emphasis. Second-order continuous quality manifests as a fearless ability to steer your own ways of doing things. It was this willingness to adjust in the face of external constraints, rather than railing against them, that allowed the architects of the Leadenhall Building in London to achieve such profound innovation in the design and construction of urban skyscrapers. Continuous quality doesn't just help organizations design their relationships with customers more effectively. It also helps them design themselves in service of those relationships.

The New QA Practitioner

QA's new role offers it the opportunity to provide deeper and greater value to digital service organizations. Succeeding in this role, however, requires a new mindset. That mindset involves lifting your gaze above the mechanics of test execution in favor of a broader, more systemic perspective. It implies becoming more curious and proactive, and engaging more deeply and openly with other disciplines.

QA engineers also need to shift their focus away from thinking of themselves as outside auditors, and toward helping service delivery teams build quality in. QA has to let go of its legacy role as a bolt-on process that happens at the end of the lifecycle. Rather than waiting for something to be delivered and pointing fingers when it isn't delivered on time or when it isn't very good when it is delivered, QA needs to become curious about the problems it sees. Why wasn't the code delivered on time? Why was the delivery so bug-ridden? Why wasn't the feedback from operations incorporated into the next release? How can we help improve the situation?

This fresh perspective leads QA to start approaching its job in terms of helping others think about quality. Given the mutual relationship between continuous design and continuous quality, this new approach is critical. QA begins asking itself questions such as:

- How can we help developers ask themselves questions about delivery timing and quality?

- How can we help operations develop skepticism regarding their own fault-tolerance designs?

- How can we help marketing and design question their own user-testing criteria?

- How can we help everyone value, seek out, and incorporate mutual input into their design and operations practices?

Acting as a chaperone or facilitator requires a very different approach from acting as a judge or arbiter. T-shaped QA engineers have knowledge about and empathy toward the other disciplines that contribute to the service delivery process. They recognize the difference between their own concern for feedback, and others' concern for action. They learn to look upon themselves as teachers and view their value based on how much others understand rather than how much they personally accomplish.

The new QA practitioner might seem nearly unrecognizable in comparison to a traditional QA engineering job description. Making the necessary mindset shift, acquiring the required new skills, and engaging in new QA practices are all very deep changes. They stand QA as we know it nearly entirely on its head.

The level of difficulty of the transition to the new QA role should not be underestimated. It reflects, though, the difficulty companies as a whole face as they try to navigate the post-industrial economy. The benefit to companies is the ability to surf and even profit from uncertainty and lack of control rather than sink under it. The benefit to QA is the chance to fundamentally transform its relationship with the rest of IT, not to mention the larger organization as a whole.

A Language for Continuous Design

The Mirror of Empathy

Continuous quality dissolves the mechanisms for achieving quality into the fabric of the organization pursuing it. It replaces the industrial, complicated-systems strategy of using QA as an enforcer with a complex-systems strategy more suited to post-industrial business. This new approach uses QA to facilitate the organization's ability to see itself from a quality-centric vantage point.

Complexity challenges us to change the way we think about the relationship between unity and diversity. Coherent systems emerge from cooperation between components that are different from one another. Practices such as service design and DevOps illustrate the need to span boundaries between disciplines. Continuous design works by engaging design and operations with each other.

Systems thinking doesn't work by mushing things together into a single amorphous blob. It defines unified wholes in terms of relationships between parts. Understanding and managing those relationships is necessary for understanding and managing the systems that arise from them.

Conversation takes place between participants that have diverse perspectives. For the conversation to be useful, the participants must be able to map their perspectives to one another. Useful conversations require empathy (the ability to see things from a viewpoint other than your own) for the very reason that their participants are not identical.

DevOps, for example, unifies development and operations into something greater. This larger system brings together two disciplines with very different starting points. Development traditionally emphasizes change, while operations traditionally emphasizes stability. In order for DevOps to succeed, the two must understand and respect each other's needs and concerns. Development must learn to value and support stability, while operations needs to learn to value and support speed.

DevOps defines a new, shared language that incorporates both worldviews. Continuous Delivery's emphasis on small batch sizes illustrates this phenomenon. Discussions of Continuous Delivery don't just stress the effect of small batch sizes on delivery speed. They also stress its effect on stability by shrinking the size and complexity of production changes.

An organization's ability to self-steer depends on its components' ability to see themselves from one another's perspectives. Autopoiesis defines a circular relationship between unity and diversity. Marketing, design, development, QA, operations, and support continuously create the digital business out of their mutual conversations. Simultaneously, though, the business guides the efforts of each discipline: the business's global customer conversations provide the context for the internal interdisciplinary conversations.

The accuracy and efficacy with which an organization self-steers depends on the clarity and efficiency with which these conversations flow. The flow of conversation depends on the ability of its participants to understand one another. In order to repeatedly change existing situations into preferred ones in an optimal way, components of a business must be able to process continuously changing signals with minimal noise or delay.

In an industrial assembly line, workers respond repeatedly to identical signals: here comes the next engine; please bolt on the same type of starter motor as you bolted onto the previous engine, in the same way, using the same wrench. The continuous design process looks very different: here comes a new customer need; it's a little different from the last one, with subtly different performance implications, and needs to be implemented in a slightly different way using a new and different application framework.

The Boundary-Spanning Mirror

Adaptive communication between disciplines, or for that matter between microservice teams, is a critical component of continuous design. It requires the ability to generate unity out of diversity: to translate between one's own and another's language, and between one's own and a shared, higher-order language. The fluidity with which the components of a digital business can effect these translations is a key determinant of quality.

Perhaps the most profound role post-industrial QA can play is to serve as a mirror that helps organizations optimize the continuous design process by clarifying translations between groups and perspectives. The concept of a *boundary-spanning mirror* comes from Guido Stompff's PhD thesis, "Facilitating Team

Cognition" (*http://bit.ly/1E7pV9g*). Stompff is a European product designer and design researcher. His thesis examined the process by which multidisciplinary product teams communicated in order to create coherent products in complicated domains such as digital printing.

Stompff concluded that designers played the role of a *boundary-spanning mirror*. They provided a mechanism by which team members from different disciplines could recognize and understand others' roles, perspectives, needs, and contributions. Designers functioned as translators between these roles and their different languages. Engineers and product managers, for example, use different terminology to refer to similar design elements. If they can't translate from their specific language to a common understanding, they won't be able to deliver a product that meets both technical and business requirements.

QA can play a similar role for digital services. QA's special role in representing the feedback component of the continuous design cycle uniquely positions it to *reflect* that cycle back at its participants. One might even claim that reflection is the heart of post-industrial QA's purpose and value.

When testing reveals a functional bug, for example, it's reflecting a translation error in the conversation between design and development. When it reveals a scalability limitation, it's reflecting a translation error between development and operations. A bug-free release doesn't just imply the absence of problems; it implies the presence of accuracy, understanding, and empathy. Just like light reflecting from a mirror, it represents a bright and positive outcome rather than merely something neutral or empty.

QA's ultimate value as a boundary-spanning mirror is to help continuous-design organizations reflect users back at themselves. Confirmation bias instinctively leads us to interpret customer feedback from our own viewpoint. Digital service organizations' deepest need for QA is the help it can provide in accurately translating another's perspective into understandable terms. That capability is the fundamental underpinning of continuous design. It is the essence of user-centeredness. Without it, digital businesses cannot hope to steer themselves in ways that genuinely help customers or that respond effectively to market disruptions.

Practicing Continuous Design

Continuous design presents a set of principles for conducting useful conversations with customers. Digital service organizations need a concrete way to put these principles into practice. In particular, they need a language through which organizational components can continuously create shared understanding in order to carry out that practice.

Such a language fills the need for a mechanism that organizations can use to understand and manipulate the relationship between diverse components and unified services. It makes continuously changing signals and the translation errors that accompany them visible and accessible. In a sense, we've come full circle, from "Does the software meet the spec?" to "Does the service meet the spec?" In the process, though, we've dramatically expanded our expectations of the kinds of requirements the spec must capture.

No longer is specification just a matter of describing expected functionality. Now the specification language must make it possible to address all four dimensions of service quality: outcomes, access, coherency, and continuity. It must be rich enough to incorporate the concerns, methods, and language of design, development, operations, and support. Even more daunting, it must provide the basis for the mirroring function through which QA helps those different perspectives translate one another's signals.

This new specification language must express the principles of continuous design. It must treat ideas such as service, feedback, failure, and learning as first-class concepts. Complexity and systems-thinking must be fundamental to its conceptual model. It must allow organizations to articulate relationships such as those between:

- Software and service
- Failure and success
- Internal and external journeys
- Design and operations

Finally and most importantly, this language needs to be a language of empathy. It must provide a way for organizations and their component parts to see, reflect upon, and correct their relationships with one another. It must provide a way to express and act to benefit others based on seeing things from others' perspectives.

In effect, what we're seeking is a programming language for continuous design. Such a language would allow digital service organizations to simultaneously describe and operate themselves as autopoietic organizations. In this way, it would resemble infrastructure as code's capacity both to document and enact system architectures. It would simultaneously function as the mirror and the reflection.

This language would be cybernetic in nature. It would treat feedback and correction of error as first-class operations. It would account for complexity and component failure. By doing so, it would provide the vocabulary for a digital service practice that unifies design, operations, and repair. Only such a practice can fully address the needs of the post-industrial economy, characterized as it is by continuously changing service relationships, complexity, and disruption.

A Brief Introduction to Promise Theory

The structures with which we manage our lives, businesses, and systems are becoming more fluid and interconnected. Complicated-systems models are no longer sufficient to help us comprehend—let alone design, build, or operate—them. We need a new modeling language that can account for the dynamic, uncertain connectedness of post-industrial business and IT. It must do so in a way that joins together not just systems but also all the disciplines, methods, and perspectives required to deliver holistic digital-service quality.

Fortunately, thanks to the work of British author, researcher, and entrepreneur Mark Burgess, a language that can satisfy such an ambitious set of requirements already exists. Burgess is the inventor of CFEngine (*http://cfengine.com*), the first infrastructure automation tool, and the progenitor of such infrastructure as code products as Chef and Puppet. He also wrote "Computer Immunology" (*http://bit.ly/1E7qqjG*), a seminal paper about self-healing distributed systems.

While working on CFEngine 3.0, Burgess was trying to find a language that could rigorously describe CFEngine's underlying operational model. This model treated complex distributed IT systems as collections of autonomous, loosely coupled, self-managing components. His research had showed him that such systems were more reliable, scalable, and resilient than systems that relied on centralized control mechanisms. Not finding a suitable existing language within IT, he formulated the basic principles of promise theory.

Burgess arrived at the concept of a *promise* as the central defining metaphor for this model. Components of a system made promises to one another. The process of making and receiving promises involved local relationships between components. Focusing on local interactions avoided problems associated with global control and knowledge mechanisms.

COMPLEX SYSTEMS AND TEENAGERS

Burgess formalized the details of promise theory into a formal mathematical model in collaboration with Jan Bergstra. Rather than trying to explain the math behind it, this chapter will illustrate promise theory with a parable. This parable starts with the claim that in order to understand complex socio-technical systems, you first need to understand teenagers.

Imagine you're hosting a party. It's an evening party. You live in Minnesota in the northernmost central United States, like I do, so it's likely to be cold out after dark. People will arrive wearing coats. You need a place to store their coats.

You'd like to store your guests' coats on the bed in your teen son's bedroom, but of course his room is a pigsty, so you need him to clean it. You ask him, "Johnny, can you please clean your room so we can use it for our party tonight?" He answers, "I promise I'll do it before dinner."

What just happened? First, he formed an intention. He voluntarily decided to do something. Second, he expressed his intention to you. If he just stood there staring at the wall thinking to himself, "I'll clean my room before dinner," that wouldn't do you any good. Third, his statement had a certain level of commitment and intensity to it. He didn't say, "Yeah, yeah, I'll do it before dinner; don't bug me." Instead, he said, "I promise." Finally, he promised to do something for your benefit. If he promised to listen to Pink Floyd in his room while you had your party, that also wouldn't do you any good.

We could define a promise as "a strongly stated intention to provide service." At this point, I need to make a small confession. Mark Burgess first introduced me to promise theory in an article he'd written titled "Promise You a Rose Garden" (*http://markburgess.org/rosegarden.pdf*). After I read it I thought, "This is lovely. It's beautiful and poetic. And I have no idea what it means." I didn't understand promise theory any better than I had before I'd read Mark's article.

After he walked me through the theory a few times, I realized I'd missed it because it was right in front of me. It was right in the words, and the words were profound. They were profound because they meant the same thing in normal usage as in technical usage, and they meant the same thing across different domains.

In order to understand promise theory, you have to understand the significance of the word *promise*. Why that word, anyway? Why not *obligation theory*, or *contract theory*, or *guarantee theory*? The thing about promises is that they aren't always kept: sometimes they are broken. Those of you who have (or have had) teenagers will understand why I chose that particular story. You know that teenagers don't always keep their promises: "I promise I'll clean my room"; "I promise I'll be home before midnight"; and so on.

We thus need to refine our definition of a promise to "a strongly stated intention to provide service, *which may or may not come to pass.*" The power of promise theory comes from the way in which it makes uncertainty explicit. We can see uncertainty in play even in apparently deterministic systems like computers. An operating system includes an API to the filesystem: open, read, write, close. It makes the following promise:

> *If you store the value 'foo' in position 23 of the file bar.txt, then come back later and ask what's in position 23 of bar.txt, the answer will be 'foo'.*

The operating system works very hard to keep this promise, using such things as parity bits, RAID, and special error detection code to help it. 999 out of 1,000 times, it keeps its promise. Every once in a while, though, it fails. It might be because someone pulled the plug in the middle of a write, or because of a bug in the software. We take our own actions because we know that failure is possible: replication, backup and restore, and so on.

THE OPPOSITE OF A PROMISE

Promise theory turns industrial control on its head. Companies typically try to accomplish complex tasks by compelling participation rather than by soliciting promises. In the language of promise theory, they create and try to enforce *obligations.*

It's easy to imagine an obligation-based version of our story. In this version, you tell Johnny that you need him to clean his room before dinner and that if he doesn't do it you'll prohibit him from going out with his friends for a week. This approach has numerous problems. First, there is still no guarantee Johnny's room will get cleaned. He might decide that asserting his independence is worth a week's grounding.

Second, just because Johnny complies with your demand doesn't mean he'll do a good job. In fact, he may do it as sloppily as he thinks he can get away with. Passive resistance is a common drawback to obligation-based control systems.

The obligation model thus fails in its primary purpose: predictability. You can't be sure whether Johnny's room will actually get cleaned, or whether it will be done well enough to use. By relying on compulsion and enforcement, you ironically compromise your ability to guarantee your own desired outcomes.

PROMISES AND TRUST

To continue our story, suppose you go out to run errands for the party. You go to the liquor store to buy some good craft beer, and stop at the supermarket for paper plates and napkins. On the way home, you call your spouse: "Hi, honey. Sorry it took so long; there was a long line in the liquor store. But I'm done now and on my way home. How are things going?" Your spouse responds with "Not very well! Johnny's sound asleep; he hasn't started on his room and won't finish in time. I'm going to have to do it, so I'm really annoyed."

At which point, you respond by saying something interesting. You don't say, "I'll kick his butt when I get home." Instead, you say, "Well, you know honey, you shouldn't be surprised. You know Johnny doesn't keep half his promises. You probably should have made a contingency plan: wake him up and remind him, or start your work early in case you need to do his, or make sure I get home early in case I need to do his."

This part of the story introduces us to the second word that's key to understanding promise theory: trust. Making uncertainty explicit puts the onus on us, as receivers of promisers, to evaluate the trustworthiness of those making them. Based on that evaluation, we need to make contingency plans. We need to keep our own promises even when promises to us our broken.

To complete the story, your spouse replies, "Yeah, I know. But you know that Johnny's been a lot better about keeping his promises lately. These days he pretty much does whatever he says he will, so I assumed I could rely on him." In other words, trust is a dynamic relationship. We don't just make one-time evaluations about things: good computer/bad computer, good teenager/bad teenager. Again, if you've had a teenager, you know that from one day to the next, you never know whether you'll get the mature near-adult or the immature near-child.

THE POWER OF PROMISES: CERTAINTY FROM UNCERTAINTY

Promise theory provides a conceptual framework that lets us model a wide range of real-world systems. It can serve as an aid to designing and operating everything from computers, to data centers, to IT organizations, to entire companies. It represents complex systems as collections of autonomous agents that voluntarily collaborate with one another by making, and sometimes keeping, promises.

Promise theory's power lies in the irony that acknowledging uncertainty allows us to achieve greater certainty. We can see this relationship at play in a variety of real-world resilient-system techniques. Familiar examples include auto-scaling load balancers, software circuit breakers, Continuous Integration, and design-for-fail.

An ecommerce website promises to return results within a certain number of milliseconds, regardless of how many customers are using the site at once. This promise is business-critical: studies have shown that even slight performance delays cost companies money in lost revenue. A site might be happily plugging along when the load balancer suddenly realizes the site is in danger of breaking its promise. In order to prevent that situation, the auto-scaler spins up another web server to handle the increased load.

The circuit-breaker pattern is a software engineering technique that illustrates the dynamic nature of trust. Imagine a software service that promises a given maximum latency in returning results. If service A depends on service B for its data, it also depends on B to maintain its latency promises. The circuit breaker pattern allows A to cut that dependency if B breaks its own latency promise.

The circuit-breaker pattern doesn't just quarantine B and forget about it. It keeps an eye on B. If B shows signs of once again being able to keep its latency promise, the circuit-breaker will provisionally reconnect it to A.

Continuous Integration represents a promise made between programmers. They promise not to check in code that breaks the build. Even experienced programmers, though, break that promise all the time. The reason development teams run Continuous Integration servers, which run unit tests on each code commit, is to catch these broken promises. Catching them quickly makes it faster and easier to fix them. As a result, downstream parts of the software delivery lifecycle see a continuous stream of kept promises.

Design-for-fail applies at all levels of digital service. It represents the understanding that each layer of a digital service must maximize its ability to keep its promises with the understanding that lower layers may fail to keep theirs. The need to design for failure applies from the infrastructure layer all the way up to UX and even beyond.

A database promises to continue operating in the presence of hardware failures. It uses clustering to help itself keep that promise. A website that mashes up a Google map promises to keep returning web pages in the presence of API failures. It uses adaptive UX design to help itself keep that promise. A hotel promises to let customers reserve rooms in the presence of website failures. It maintains a phone number for call-in reservations to help itself keep that promise.

THE POWER OF PROMISES: SCALABILITY AND RESILIENCE

Industrial control models treat systems as being composed of components that are tightly locked together. The components behave correctly because the system forces them to do so. They system in turn behaves by forcing its expectations upon its components.

Centralized control maximizes global efficiency. At the same time, however, it compromises scalability and resilience. The larger the system, the harder it becomes to consistently propagate control signals. Tight coupling also makes large systems more susceptible to failure given the greater number of failure-prone components.

Complex systems rely on local rather than global control (e.g., birds coordinating with their immediate neighbors). Their emergent structures are resilient to component failures. Promise theory takes a post-industrial, complex-systems worldview. This worldview makes promise theory a powerful tool for designing, building, and operating systems that are scalable and resilient.

Promises encode local control and communication between components. The nature of a promise is in the eyes of its creator and its beholder, not the system as a whole. An application must judge the promise made to it by a database, regardless of the promise made to the database by the hardware. The fact that the application can't judge the hardware's promise also means that it doesn't have to.

Microservices, especially when used as an organizational and not just architectural pattern, capture this quality of promises. Microservices make promises to one another regarding functionality, quality, and integrity in the face of change. They are free to pursue their own internal processes as long as they keep their external promises. This freedom gives each component greater flexibility and prevents control signals from having to propagate across the entire system or organization.

The Agile movement talks a great deal about self-organizing teams. In nature, systems don't self-organize in order to do what they're told. They organize to fulfill some internal purpose. A development team that makes voluntary promises to fulfill customers' needs is more powerful, flexible, and scalable than one that passively responds to externally imposed requirements.

Agile is a wonderful strategy for responding to uncertain, changing environments. When we try to scale it up from individual teams to entire IT organizations, we seem to forget everything we learned. We suddenly abandon our complex-systems model in the belief that coherent systems can only result from complicated-systems control. "Well, this sloppy Agile stuff is all well and good, but we have to deliver complete features, and features cross teams, and we need to be sure every team delivers the right thing on time."

Agile can learn a great deal from the housing industry in this regard. The construction process is rife with uncertainty. The painter falls off a ladder, breaks his leg, and can't paint for two weeks: what do you do? There's an overseas dock strike and the teak delivery is delayed or the price goes up: what do you do? Regardless of failures at the subproject level, the architect and general contractor still have to build—on time and on budget—a house that the client can live in.

Contractors and subcontractors make promises. These promises are local. The client doesn't care what happened to the painter. The general contractor, in turn, doesn't care whether the painter can get paint from his favorite store. Keeping the making and evaluating of promises local avoids the fragility and paralysis that comes from overlarge, centralized control mechanisms.

Unifying the Mirror and the Reflection

Promise theory offers a concrete language for modeling the kinds of complex socio-technical systems that make up infused digital services. It can be used to represent relationships at all levels of a service: from hardware, to software, to system architecture, to human interactions, to organizational structures. Just as databases make promises, so too do customer service agents and Agile development teams.

Promise theory unifies the boundary-spanning mirror and its reflection by unifying design, operations, and validation. Design becomes a matter of determining what promises to make and how to keep them. Operations becomes a matter of continually striving to keep your promises even when promises made to you are broken. Validation (aka feedback) becomes a matter of continually testing the gap between design and operations. It is in this sense that promise theory transforms "validating the software against the spec" to "validating the service against the spec."

Unifying the mirror with the reflection naturally leads to continuous design. The definition of a promise as "a strongly stated intention to provide service that may or may not come to pass" captures the essences of the core continuous design principles. A promise is a commitment to do something to serve others. The possibility of failure implies the need to use feedback to detect and correct failures. The incorporation of failure within the definition of success, along with the understanding that promise-keeping requires continual detection and repair, drives operations as learning.

Promise theory is inherently cybernetic. It defines a domain-independent operating model that strives to achieve certainty—that is, control—by conversing with uncertainty. Its understanding of promise-keeping as a never-ending process of trust evaluation and promise repair implies the need to use operations as an input to design. The same mechanism that maintains control also provides the information needed to improve your own control mechanisms.

Promise theory thus defines a language for practicing continuous design. It offers a concrete method for thinking about, designing, and operating responsive digital services. It provides a unified framework for addressing the principles of continuous design and validating their effectiveness across the four dimensions of digital service.

Service as a Chain of Promises

Imagine you're a weary traveler. You've just arrived at your hotel and you're not in a good mood. Your flight was delayed; there was terrible turbulence when it finally got under way; the cab driver on the way from the airport was crazy.

You finally make it to your hotel and present yourself at the front desk. It's late, you're tired, and you just want to check in and get to your room. The desk agent greets you with a sad look and informs you in an apologetic voice that he can't check you in because the check-in system is down.

What just happened is a service failure. To understand the true nature of the failure and to understand how best to repair it, we need to revisit the definition of service. The language of promise theory can help make sense of the problem and guide the discovery of an effective solution.

The Service Promise

For the purpose of this analysis, we can define service along three axes:

- Experience, not thing
- Relationship, not transaction
- Co-creation, not delivery

Customers experience service as something that unfolds over time across touchpoints, not as a momentary thing. Your experience of a hotel begins when you log onto its website to reserve a room. The site is slow, hard to use, and doesn't let you search for rooms based on your needs. You manage to book a room, arrive at the hotel, and can't check in because the systems are down. You finally check in, find your room, and turn on the TV to find a lousy channel selection. The WiFi network is so slow you can barely use it. You can't sleep because someone is using the ice machine next to your room all night. By the time you check out the next morning, you can't wait to leave!

Customers judge service quality based on an ongoing relationship, not just a single transaction. The majority of Twitter posts related to airline complaints don't say "my flight was delayed." Instead, people tweet complaints like "As usual, my flight was delayed," or "News flash! For once my flight wasn't delayed!"

Finally, and most subtly but most importantly, service doesn't pour value into a product and hand it to a customer in exchange for money. The fact that you own a hotel and sold someone a room in it doesn't provide value by itself. The guest doesn't gain benefit from her room until she actually stays in it. The value comes from her interaction with you throughout the lifecycle of the customer journey. If she's had a lousy flight, her mood contributes to the experience.

In order to generate customer satisfaction by co-creating value with your guest, you have to account for that mood. As a service organization, you make a promise to your customers to help them accomplish their jobs-to-be-done. You make that promise in the context of the entirety of the service experience and the entirety of the customer relationship.

As Rebecca Sinclair, a designer from Airbnb, said during a talk about the company's site redesign process, "We realized that the product was the trip." What she meant by that comment is that people don't reserve a hotel room because they want a hotel room. They do it because they need a place to stay while they're traveling.

The purpose of the hotel is to help them have a satisfying trip. The service organization needs to understand and account for the entire context that surrounds the hotel stay. In other words, the service organization needs to make the right promises.

Having made the right customer promise, the entire organization must work together to fulfill it. Check-in software, front-desk agents, and maids all contribute to the success or failure of the customer's service experience. As autonomous components of a complex socio-technical system, they all make promises to the customer and to one another. Some of those promises may be as simple as promising to be patient with someone who's tired and cranky.

MAKING THE RIGHT PROMISE

What is the traveler's job-to-be-done upon arrival at the hotel check-in desk? Is it to get entered into the computer, pay for the room, and retrieve the room key? Not really. The traveler's real job-to-be-done is to make the transition from travel and stress to rest and relaxation.

We all intuitively understand what this means. Checking in is the final psychological gate to cross before you can take a break from traveling. As you enter the lobby and look for the check-in desk, you can feel that last bit of anxiety before you can relax and stop thinking, "Which concourse does my flight leave from? Where do you go to get a taxi in this airport? What's the hotel's address? How long til I get there?"

To return to the question that opened this chapter: when the desk agent says, "I can't check you in because the system is down," what is the nature of the service failure? The failure consists of the fact that the hotel has made the wrong promise. What the hotel has promised is merely that the desk agent will accurately translate the words from the guest's mouth into commands the computer can understand.

What guests really need is a promise to help them make the transition from travel and stress to rest and relaxation. By passively stating, "I can't check you in," the desk agent has abdicated his human autonomy. Were he able to understand and act upon the correct promise, he might be able to keep that promise even in the face of a software failure.

To account for the possibility of a reservation system outage, the hotel might set aside a few rooms to be tracked using paper and pencil. Alternatively, desk agents might be trained to do something even simpler: they could just tell the guest, "I'm sorry I can't check you in right now because our system is down. Why don't you leave your luggage with me, and go over to the hotel bar and have a complimentary drink and some hors d'oeuvres? I'll come get you when I'm able to check you in."

In the latter scenario, the hotel still can't check the guest in. The guest still can't get to her room. She can, however, let go of her luggage and relax. The difference in customer satisfaction between "I can't check you in" and "I can't check you in but I can offer you a drink and a chance to sit down" is tremendous. By making the right promise, the hotel enables its human components to act autonomously for the customers' benefit.

The Chain of Promises

A service is a complex chain of promises. We can see this characteristic at play in our earlier story about hosting a party. What is the fundamental promise one makes as a host? It's to help your friends have a good time on a Friday night—forget their troubles, catch up with friends, enjoy good food and drink. In order to keep this promise, you need to keep a chain of other promises. You have to cook food and clean the house, Johnny needs to clean his room, and either you or your spouse needs to do the shopping and get some good beer.

The promise to serve good beer is where things get especially interesting, and where the concept of service systems from service-dominant logic comes into play. In order for you to be able to serve good beer, there must be a suitable liquor store nearby. The store owner must make and keep a promise to run a profitable business; otherwise, the store won't be there when the time comes for your party. The store owner must also have good taste in order to keep the promise to stock good quality beer.

The chain of promises needed to deliver service quality starts with the customer. It expresses the service organization's internal structure, and expands from there to encompass the relationship between the organization and the customer. Not content to stop at that level, it continues on to integrate other service organizations.

Promise theory's greatest power comes from its ability to facilitate systems thinking. Promises inherently involve relationships: in this case, between promisers and promisees. By focusing our attention on relationships, promise theory leads our minds up and out toward systems. This holistic, expansive momentum naturally illuminates all the internal and external components that contribute to value co-creation.

Using Promises to Achieve Digital Service Quality

Promise theory provides a set of questions organizations and their component parts can use to help themselves think holistically about complex, resilient service systems across dimensions and domains. These questions include:

- Which promises should we make?
- What promises do we need from others in order to keep our own promises?

- How do we maximize our trustworthiness by keeping the promises we make and repairing the ones we break?

- What promises do our customers need to make?

MAKING PROMISES

Software development experts often say that it doesn't matter how good your code is if it's not the right code. In other words, if the software you build doesn't meet real customer needs, then it doesn't matter how well constructed it is. "Which promises should we make?" is the service version of the same question. The hotel failed the customer, not because it broke its promise to keep the check-in system online, but because that wasn't the right promise to make in the first place.

Digital services rely on collaboration between diverse components. Front-office software must integrate with back-office software. IT systems must integrate with human operators. Marketers and designers must collaborate with developers and system administrators.

In order to contribute to overall service quality by doing its particular job, each component of a digital service must in turn acquire services from other components. It must do so in a way that doesn't compromise resilience or scalability. Rather than relying on top-down, industrial-style control signals to tell it what to expect from its neighbors, it must, like the flocking bird, "calibrate its own flight path."

Asking itself what promises it needs from others to keep its own promises guides each service component in accurately identifying what help it needs and from whom. The phrasing "in order to keep my own promises" helps each component evaluate that help in terms of its own empathic purpose. It keeps the ongoing assessment of quality focused on the achievement of internally generated goals.

KEEPING PROMISES

Knowing which promises to make and which promises to seek from others is necessary but not sufficient. What matters is keeping promises, not making them. Whether others keep their promises to you is of little interest to the recipients of your promises. Approaching service design and operations from the perspective of continually ensuring your ability to keep promises in the face of uncertainty is the heart of promise theory's contribution to service quality.

Of course, in a complex system, perfect promise-keeping is infeasible. Component failures are inevitable. No real-world service can guarantee that none of those failures will become visible to its customers. Along with evaluating its ability to keep its promises, services and service components must evaluate and design for rapid and customer-centric promise repair. Maximizing quality includes not just preventing defects but also fixing them.

HELPING OTHERS KEEP THEIR PROMISES

The final question, "What promises do our customers need to make?", really should be the one that drives all others. In the language of promise theory, a job-to-be-done is a promise-to-be-kept. The hotel guest needs a good night's sleep because he's promised to give an inspiring talk on promise theory at a conference the next morning. Or perhaps he's promised to show his kids a good time at Disneyland. Or maybe he's promised to make his quarterly sales quota by closing a big deal in a meeting the next day with a prospective customer.

Regardless of the scenario, this question drives the entire organization and each of its component parts from the proper definition of service quality. It frames all the other promise-keeping questions in terms of empathy. It leads me to judge my success as a service provider not in terms of what I accomplish but in terms of what I enable others to accomplish.

Maintaining the Illusion of Continuous Quality

You might think it's possible to achieve continuous quality by keeping your promises. In reality, there is no such thing as truly continuous quality. Failure is inevitable in complex systems. Attempts to guarantee the invisibility of those failures ironically degrade quality by transforming sloppy resilience into hard brittleness.

Even beyond the level of explicit defects, quality is inherently discontinuous. Regardless of how perfectly a service meets a customer's needs, those needs will change. Simply by introducing new functionality, service changes user behavior and thus the design context. Continuous design arises in response to the fact that solving existing problems creates new ones.

In the post-industrial economy, achieving continual quality really entails achieving continual repair. Continual repair operates on multiple levels:

- Keeping the promises you've made by repairing the failures that inevitably plague them

- Improving your ability to keep the promises you've made by evolving your existing repair mechanisms

- Improving the quality of your service as a whole by repairing failures due to making the wrong promises

Continuous design is a cross-functional method for continual repair. Continuous design principles such as maximizing feedback, designing for failure, operating to learn, and seeking empathy account for failure on multiple levels. Its unique value, though, comes from its ability to address failures due to making the wrong promises.

As service and customer needs evolve in response to each other, their mutual fitness is subject to continual failures. By adding an expense reporting feature, for example, an invoicing service leads its users to think beyond purely invoicing. The promise to help them track their time suddenly becomes insufficient. Using operations as input to design exposes this new "defect." Treating value propositions as promises lets service providers dynamically keep and improve them in the same was as they keep application and infrastructure promises.

Promise theory provides a coherent, unified language for practicing continuous design at all levels of digital service, from data-center management to business model design. It also lets organizations use the same language for internal self-design. Promise theory models systems in domain-independent language. Nothing about it is specific to IT, design, or any other discipline. It creates a common language that all the disciplines required for digital service can use to understand and manage their relationships with one another.

Digital service depends on frictionless collaboration between diverse components. Effective collaboration involves not just understanding but also the ability to navigate misunderstanding. The mirror of empathy is needed because different parts of an organization have different perspectives. Promise theory's incorporation of failure and repair aids collaboration by accounting for the need to repair misunderstanding as an integral part of ongoing service operations.

Continuous design is a strategy by which IT can respond to the arrival of the post-industrial economy, characterized as it is by complex, disruptive, digitally infused service. Promise theory captures the essence of post-industrial service.

One could define *service* as "a strongly stated intention to provide benefit." Service is continual; its quality is judged and must be continually maintained. The fact that quality in the form of benefit "may or may not come to pass" keeps service organizations focused on customer empathy as something to achieve every day through continuous self-redesign, not something to put in place and then just run like a factory.

12

Promising a Good Night's Sleep: The Digitally Infused Hotel

Hotels are services with which we all are familiar. The hotel industry makes an ideal case study for illustrating the application of promise theory to real-world businesses. At first blush, a hotel is a very physical service. It involves buildings, rooms, elevators, lobbies, desk agents, and maid service. We think of a hotel as being about people and architecture. Now, though, we're entering the age when you can use your watch to unlock your hotel room door. Hotels thus are becoming deeply digitally infused.

Even the watch itself is an infused object. Traditionally, watches are very physical things. They're heavy; you wrap them around your wrist; they're full of gears and make ticking noises. Now, though, watches have become computers.

Infusion impacts the hotel experience even without *Dick Tracy*–style watches. We take it for granted that we can reserve a room over the Internet. We expect to be able to check out using the television in our room. Promise theory's domain-independence makes it as useful for modeling services—such as hotels, which cross human, physical, and technical boundaries—as for modeling pure software services.

Infusion is not just superficial. It pervades all levels of a service such as a hotel. We can see this phenomenon at work by revisiting the hotel scenario in more detail. As we do, we'll notice that its promises don't necessarily break out along traditional organizational silos, nor along divisions between the four dimensions of service: outcomes, access, coherency, and adaptability. In fact, they don't even break out separately between physical and virtual realms.

Each of these concerns infuses the others at various levels. Service design and DevOps reflect the need to model infusion in their attempts to bridge differences between cultures, practices, and channels. As we will see in the hotel scenario, we can use promises to model the entirety of the relationships necessary to design and operate an infused service.

Promising Rest

What is the core promise that a hotel operator makes to its guests?

I promise to help you rest as part of a trip.

In order to keep that promise, the hotel first must help a potential guest find and reserve a satisfactory room. The chain of promises, like the customer journey, begins prior to the moment someone actually becomes a customer. It starts with her first interaction of any kind with the service.

This could go as far back as the process by which a potential guest discovers the hotel in the first place. It's at that point that the hotel makes its first promise. There's a reason people say things like, "This looks promising."

However people discover a promising hotel, these days they generally book a room via its website. The reservation site makes a series of promises. These promises cross usability, functionality, and operations domains. For example:

I promise to help you find a satisfactory room.

I promise to let you navigate by meaningful search criteria.

I promise to show you the cost of the room before you reserve it.

I promise not to crash in the middle of a transaction.

I promise not to lose your reservation prior to your arrival.

I promise not to let anyone steal your credit card number.

It's important to note that people don't want to reserve just any random room. They might have specific needs in order to get a good night's sleep: a king-size bed if the guest is tall; a low floor if he's afraid of heights; a nonsmoking room if she has asthma. In order to fulfill its core promise to help guests rest, the hotel needs to let guests reserve rooms that meet those criteria.

The reservation system's promises reflect the need to provide inseparable functionality, usability, and operability. Actions such as searching and paying for a room need to work properly. They need to be accessible so that users can accomplish them without confusion. At the same time, however, the reservation system will sabotage its usefulness if it doesn't also keep its promises of availability, data consistency, and security.

PROMISING THE TRANSITION TO REST

Having successfully booked a room, at some point the guest arrives at the hotel. The hotel's next promise, as previously discussed, is:

> *I promise to help you make the transition from travel and stress to rest and relaxation.*

If you're like me and you find it challenging to navigate strange places at night after getting off a plane, the simple task of finding your way through the lobby to the front desk can be intimidating. If the lobby is confusing or the check-in desk is hidden, the hotel has in effect broken an interior design micro-promise.

Eventually you find your way to the check-in desk and encounter a desk agent. The agent's fundamental promise is to be friendly, helpful, and autonomous. Agents are people and implicitly promise to behave that way. This promise is sadly not as obvious as it should be in this age where everything is computer-mediated. To put it a bit cynically:

> *I promise to do more than just type your name and credit card number into the computer.*

Again, as previously discussed, the agent's job is to help you transition from stress to relaxation regardless of the state of the check-in system.

PROMISING TO HELP EMPLOYEES

The desk agent's promise doesn't let the check-in system off the hook. It makes a similar set of promises to those made to customers by the reservation system. Value co-creation implies the empowerment of employees as well as customers. The newly popular notion of *IT-as-a-Service* means treating internal IT users as first-class rather than second-class citizens.

The desk agent's goal is to help newly arrived guests transition from stress to relaxation. The most straightforward way to do that is to get them checked in to their rooms. The check-in system makes a promise to that effect:

I promise to help you get guests checked in.

To that end, the check-in system must make its own promises of usability, functionality, and operability. Promise theory's support for generating certainty from uncertainty relies on a corresponding set of commitments. On the one hand, the promisee commits to providing benefit regardless of the promiser's performance. On the other hand, the promiser commits to doing his best to keep his promise.

In complex system, everyone is simultaneously a promiser and a promisee. In plain language, service quality relies on everyone doing their best. Internal IT systems are no exception. In the case of the check-in system:

I promise to help you efficiently navigate the check-in process.

I promise to interface with the reservation system to quickly retrieve the guest's information.

I promise not to crash in the middle of a transaction.

I promise not to let anyone steal the guest's credit card number.

PROMISING SERVICE INTEGRATION

From the perspective of service-dominant logic, a hotel co-creates value with its guests as part of larger service systems. The guest may be traveling as an employee of a company. She may be traveling in order to meet with a customer or potential customer. As part of her trip, she may need the help of airlines and taxi companies as well as the hotel.

The hotel thus makes promises related to the larger service context that surrounds it and the guest:

I promise to update your frequent flier miles with your airline based on your hotel stay.

I promise to help you find a nice restaurant to entertain your customer.

I promise to help you find a ride to the airport in the morning.

I promise to help you connect to your corporate network in order to file a trip report.

Phrasing the need for hotel-room Internet access as a promise helps understand why guests get so frustrated with slow hotel WiFi networks. Guests need

WiFi access because they have work to do. Their hotel room and their hotel stay are just part of a larger set of service integration activities. Anything that increases stress compromises the hotel's ability to keep its core promise.

Keeping and Repairing Promises

Identifying which promises to make is merely the first step in the promise design process. Next, each service must identify how to maximize its ability to keep them. In the case of the promise to help a potential guest find a satisfactory room, for example, the hotel might support over-the-phone reservations for cases where the reservation website is offline. It might empower its desk agents to offer guests a free drink if they can't check them in. It might conscript someone from the bar to help check in guests when there's a long line at the front desk.

Because perfect promise keeping, even with contingencies, is infeasible, a service also needs to account for the possibility of failure and determine how best to repair broken promises. Security breaches are great ways to see how well a company understands this reality. Security promises have more emotional impact than most: a reservation system outage is an inconvenience, whereas theft of private information feels like an invasion.

Companies have opportunities to reinforce or degrade their brands based on how they respond to breaking their security promises. Zappos' handling of its 2012 security breach (as mentioned in Chapter 6) illustrates the power of thinking in promises. The entire tone of its post-breach communications reflected the sentiment that "we broke our promise to keep your information safe." The company took responsibility for the need to repair customers' trust. As a result of this attitude, the broken promise ironically increased the perceived quality of the Zappos brand.

KEEPING AND REPAIRING FUNCTIONALITY PROMISES

Among the promises IT systems make are those related to functionality. Like any other component of an infused service, development teams need strategies for keeping these promises. Continuous Integration is such a strategy. The development team promises to deliver code that doesn't break features that used to work. Running automated regression tests as part of continuous builds helps keep that promise.

Development teams also promise to deliver code that is genuinely useful. Sprint demos help keep that promise. The purpose of a sprint demo is not to show people what the team is about to release. Instead, it provides an opportunity to validate the usefulness of the code that's been written and to adjust it as needed.

Continuous Delivery bills itself as a way to increase velocity. Its greatest benefit, though, is as a repair mechanism. The ability to move a one-line bug fix all the way through the development/delivery lifecycle is where Continuous Delivery really shines. In other words, with a Continuous Delivery process in place, a development team can very quickly repair its broken promise to deliver working code.

KEEPING AND REPAIRING OPERABILITY PROMISES

IT organizations go to great lengths to keep their operability promises. Databases would be much less complicated or expensive without support for resilience mechanisms such as clustering, replication, and failover. Networks and hardware would be simpler and cheaper without built-in redundancy. Data centers would be drastically less interesting-looking without wall-sized batteries, sophisticated fire suppression systems, or industrial-strength diesel generators.

Fault-tolerance strives to keep failures invisible to human operators. Like any other promise, it is an imperfect strategy. IT-level failures don't, however, inevitably surface to become visible to customers. Monitoring is a fundamental operations strategy for keeping customer promises in the face of a broken infrastructure promise. By proactively alerting operators of broken infrastructure-level promises, it provides a window of opportunity for repairing promises at the infrastructure level in a way that maintains their invisibility to customers.

Of course, even monitoring is an imperfect promise-keeping strategy. Just as a company as a whole must account for outages, so must IT. Repairing broken operability promises involves more than just fixing broken systems. It also involves internal and external communications. IT's ability to communicate during and after an outage impacts its perceived trustworthiness within the company, as well as the company's perceived trustworthiness with its customers.

Fixing broken system promises and communicating skillfully about them is necessary but not sufficient for IT-as-a-Service. Repeatedly breaking the same promises degrades trust, regardless of how well one repairs them. IT also needs to treat outages as learning opportunities. Customers don't evaluate service adaptability purely from a functional perspective. They also evaluate it in terms of operability.

The more pervasive digital infusion becomes, the more likely it will be that ordinary consumers begin to understand the inevitability of failure. When every service provider experiences outages, consumers differentiate them by their ability to learn from failure. Providers that don't make the same mistakes twice gain trust.

13

Brands as Promise-Marks

So far, the hotel scenario has described how the service operator wants things to be. It has used promises to design the desired customer experience at any given point in time. Continuous design also necessitates considering how the operator wants things to become. To understand what this notion means, we need to take a step back and revisit the concepts of brand and brand quality using the language of promises.

The etymology of the word *brand* dates back to its original meaning: an identifying mark made by a hot iron. Traditionally, marketers used messaging to leave a mark on a customer. That mark would trigger desired impressions in the minds of customers. When people thought of a BMW, they would associate it with excitement. When they thought of a Mercedes-Benz, they would associate it with elegance. When they thought of a Honda, they would associate it with practicality.

By shifting the dynamic through which brand messages are generated, new technologies such as social media are making brand maintenance more interactive. As a result, marketers now often talk about consumers "engaging with brands." Something about that phrase, though, seems odd. If the idea of brand is as intangible as marketers and designers claim, then how can you "engage" with it? It doesn't seem as if a brand can do anything to anyone, or that one can do anything to a brand. It would seem more accurate to think of a brand as the result of a company's and its customers' mutual actions.

Peter Laudenslager (*http://linkd.in/1E7sQP*), a cloud computing strategy specialist, remarked that "a brand is a package of assumptions/expectations... what people assume and expect, true or no, intentional or no." (*http://bit.ly/1E7sVlS*) This definition seems sensible. But where do the particular assumptions/expectations associated with a brand come from? Customers don't make them up out of thin air. People don't expect an airline to help them lose weight. They do expect it to get them to their destinations on time with minimum hassle.

The package of assumptions and expectations one has about a company starts with the set of explicit and implicit promises the company makes. An airline promises to transport you from one city to another within a reasonable time frame. It promises to feed you along the way and to get your luggage to the same location at the same time as your arrival. Most importantly, regardless of how conveniently it does so, the airline promises to deliver you safely.

The Cybernetic Brand

Customers' assumptions and expectations evolve based on their experience of how well a company keeps its promises. Travelers may come to assume, for example, that a particular airline won't get them to their destination on time and that they'll do it with maximum rather than minimum hassle. We can thus define a brand as "a package of promises made, kept, and broken."

A brand is essentially the mark of a company's trustworthiness based on the history of customers' interactions with it. Promise theory teaches us that trust is dynamic. A service's ability to keep its promises and make the right ones can change over time. Maintaining brand quality shifts from crafting and targeting messages to validating the quality of the brand's promises.

The post-industrial economy is transforming brand maintenance from an activity based on broadcasting messages to one based on conducting digital conversations. Promise-based service design needs to address this new, conversational nature. It must define the mechanisms by which organizations can use feedback to steer their own promises. Given the autopoietic basis of self-steering, it must do so in a way that captures the dynamic, circular relationships between services and their components.

From this perspective, the most fundamental promise a service makes is:

I promise to conduct useful customer conversations.

In order to keep that promise:

I promise to continually deliver improved capability.

I promise to listen to customers' responses.

I promise to guide future actions by those responses.

Promising to Be a Conversational Medium

Defining a service organization in this way essentially defines it as a cybernetic system. Doing so has profound implications for IT. It radically transforms IT's purpose from "maintaining stable information systems" to "enabling useful customer conversations." In fact, IT's fundamental promise becomes neither more nor less than:

> *I promise to enable useful customer conversations.*

This new promise illuminates the forces behind the need for IT transformation and provides the rationale for continuous design. This rationale might look eerily familiar:

> *I promise to design for service, not just software.*
>
> *I promise to minimize latency and maximize feedback.*
>
> *I promise to design for failure and operate to learn.*
>
> *I promise to use operations as input to design.*
>
> *I promise to seek empathy.*

Making New Promises

By embarking on the practice of continuous design, IT makes a new set of promises to the organization it serves. In order for IT to keep these new promises, its component parts also need to undergo their own transformations. Using promises to define the digital conversational medium is more than just a convenience or parlor trick. When a design group adopts a leaner approach to producing UX solutions, or a development group adopts a more iterative approach to producing code, or an operations group adopts a more collaborative approach to managing production changes, they are changing the promises they make. In other words, they are voluntarily committing to changing the way they work for others' benefit.

The Agile Manifesto defined a cultural transformation in software development by making a set of value statements. The heart of this transformation was contained in the statement that "we value responding to change over following a plan." In other words, "We believe it's more important to promise flexibility than to promise foresight."

The point is that the Manifesto's proposed transformation relies on its adherents committing to a new belief, not just following a new procedure. Promising to value responding to change over following a plan does not mean promising to show up in a meeting room every morning at 9:00 AM and stand in that room for 15 minutes and answer three questions about the work one intends to do that day. It means showing up in the room and answering the questions because the team recognizes that doing so is an effective way to continuously adapt to changing customer needs.

MAKING NEW PROMISES TO ONE ANOTHER

IT transformation triggers, as well as arises from, the transformation of its internal relationships. In addition to making new promises to customers and to the business as a whole, IT's component parts also need to make new promises to one another. The core promise that makes it possible for marketing, design, development, QA, operations, and support to cohere is:

I promise to value systems over components.

This promise guides the entirety of the activity within a digital service. It could even be argued that it defines the essence of post-industrial business. Service means seeing companies and customers as part of a larger whole. Infusion means seeing systems and people, software and hardware, as part of a larger whole. Disruption means seeing marketing and design (what comes next) and operations (what comes now) as part of a larger whole.

In order to keep this promise, continuous-design practitioners must think across disciplines while doing their own work. They must make specific conversational promises to one another. A UX designer, for example, might make the following commitments:

I promise to design interfaces that gracefully respond to API failures.

I promise to design interfaces that minimize browser resource requirements.

I promise to design digital interfaces that are compatible with physical interfaces.

A developer might make the following commitments:

I promise to deliver code that is scalable.

I promise to deliver code that is fault-tolerant.

I promise to deliver code that is secure.

I promise to deliver code that provides meaningful metrics.

An operations engineer might make the following commitments:

I promise to minimize deployment friction.

I promise to provide visibility to meaningful metrics.

I promise to help design and development deliver operational quality.

KEEPING AND REPAIRING CONVERSATIONAL PROMISES

As with any other kind of promise, the promises that drive continuous design are subject to failure. They require attention and mechanisms to help keep and repair them. Fortunately, many of the techniques that define practices such as Agile and DevOps serve exactly that purpose.

Traditionally, IT organizations used heavyweight governance procedures to ensure internal promise-keeping. InfoSec teams, for example, would execute lengthy security reviews at the end of development cycles just prior to their release. This approach placed InfoSec in a similar position as QA. Their relationship with marketing and development was antagonistic and wasteful.

What's needed, though, is a post-industrial rather than an industrial solution, one that respects and leverages the autonomy and empathy of its participants. All of the practices that make up the conversational medium rely on intimate cross-functional collaboration. Taken together, they enable collaboration throughout the service organization.

Cross-functional teams create the environment that enables the various disciplines to keep their mutual promises. Cross-functional design spikes, for example, provide opportunities for designers, developers, and ops engineers to expose gaps in one another's promises: "That's a nice home page design, but it won't work in mobile browsers" or "That's a good way to make the network more secure, but it will double the time it takes to deploy new code."

Done properly, CI uses automated tests to reify the process of intimate cross-functional review. A comprehensive test suite includes performance, resilience, and compatibility tests. Running this test suite regularly gives team members continuous visibility into broken mutual promises.

Intimate collaboration requires trust. Cross-functional design reviews aren't helpful if people see them as venues for criticism rather than assistance. Trust requires understanding and empathy.

Shared tools can help teams improve understanding and empathy by creating common language and experience. If everyone from marketing to operations uses a kanban board,[1] for example, they all have something in common. That commonality makes it easier for them to help each other keep mutual promises.

MICROSERVICES AS CONVERSATIONAL PROMISES

To succeed, digital service must bind together disciplines across the technical and nontechnical spectrum into coherent whole systems. At the same time, self-steering requires the ability for the components of those systems to flex with respect to one another. An organization based on cross-functional microservices (or as Dave Gray called it in his book *The Connected Company*, the *podular* organization) is the ultimate expression of this principle.

Microservice pods have the potential to bring continuous design's promises to life. When done right, microservices represent a full expression of the view of systems as collections of autonomous components that voluntarily cooperate to provide adaptive service. For this vision to succeed, however, it needs to take the word *service* seriously.

Microservices gain their flexibility and resilience through loose coupling. Each pod has the freedom to pursue its own release schedule and operational model. That freedom is what allows agility to scale without becoming brittle.

With freedom comes responsibility. "Do what thou wilt" cannot be the whole of the law. Instead, it must be "Do what thou wilt for the benefit of others." Everything we've said so far about service applies to microservice pods. To put it more formally, each pod makes the following commitment:

I promise to treat other microservices as service customers.

In other words, each microservice pod promises to continually design itself. First and foremost, it promises to conceive of itself as a service organization, not just a software component. That promise means helping other pods understand how this microservice works. It means supporting them in using it. It means presenting stable interfaces and evolving those interfaces in digestible ways.

---

1 A kanban board is a shared system for visualizing and controlling the flow of work through a process or team.

The promise to design for service, not just software, also applies to each microservice's internal relationships. The members of microservice pods need to make the same promises to one another that designers, developers, and operations staff make to one another in general. The way in which microservices bind together functionality and operability into complex systems built from small components, for example, implies the need for DevOps (*http://bit.ly/1E7tjAQ*).

Microservices promise to contribute to global continuous design by minimizing latency and maximizing feedback. They promise to design for failure and operate to learn. Optimizing the ability to keep these promises is a key part of the rationale for adopting microservices architectures in the first place.

Microservice-based architectures and organizations can help IT maximize its continuous design capabilities. In order to truly drive conversational quality, however, they must infuse themselves with the same mindset to which they claim to contribute. In order for operations to provide input to design, that input needs to flow fluidly throughout IT. Each microservice must empathize with its dependents. It must listen to their feedback, and view continual adaptation to that feedback as its core mission.

Promising Continuous Design

A digital service's brand quality relies on its ability to make and keep promises over time. Unfortunately, complexity conspires against an infused organization's promise-keeping efforts. It constantly generates failure in ways against which no organization can perfectly defend.

These failures arise on multiple levels. Some of them take the form of defects. Some of them take the form of missing features. Some of them take the form of completely unforeseen needs generated by the larger context.

Regardless of the cause, digital services are doomed to continually confront their customers with failures of various sorts. Consumers are coming to expect this eventuality. As a result, they begin to judge brand quality not just on a company's ability to prevent failure but on its ability to learn from it. This reality makes continuous design indispensable to post-industrial businesses. All the promises they make and everything they do to keep those promises must happen in service to the fundamental promise to learn, improve, and evolve on their customers' behalf.

A company that merely apologizes over and over again without proving that it is improving will eventually degrade its brand. Digital service customers intuitively grasp the cybernetic nature of post-industrial business. They praise companies that solve problems, whether those problems be incomplete feature sets, confusing user interfaces, or service outages. Conversely, they punish companies that make empty promises. In this way, many customers are ahead of their vendors in understanding the promissory nature of the digital brand conversation.

14

Thinking in Promises

Promise theory brings the heady concepts of cybernetics down to earth. It turns a deep, abstract conceptual framework into a practical methodology post-industrial IT can use to transform its role in twenty-first-century business. It provides a comprehensive continuous design language that digital service organizations can use to steer themselves by conducting useful digital brand conversations.

The goal of this book is to help readers learn to *think in promises*. Promise thinking, like design thinking, is a way of doing, not just a way of contemplating or analyzing. Some critics of design thinking have identified a supposed need for *design doing*. In reality, though, design thinking is already a method for designing things (whether those things be annual reports, buildings, taxi services, or government policies) from a certain perspective or way of thinking. The design-thinking perspective is user-centered, iterative, and feedback-driven.

Similarly, promise thinking is a method for designing and operating complex socio-technical systems. The promise thinking perspective sees systems as composed of autonomous, user-centered, feedback-driven components. These components approach *control*—that is, continuous quality—as a process of continual repair. That approach helps organizations achieve the responsiveness required to succeed in an economy increasingly characterized by service, infusion, complexity, and disruption.

Doing Continuous Design

As we have seen, promise theory expresses the principles of continuous design: service, feedback, learning, continuity, and empathy. Promise thinking shifts the practitioners' focus away from the products being made and toward creating benefit for others by dynamically maintaining trustworthiness. By valuing conversations over solutions, it naturally leads to continuous design.

Promise thinking encourages systems thinking. By focusing on commitments made for others' benefit, it frames design in terms of relationships. It points the mind upward and outward. Designers and operators alike begin to see entire digital service organizations as chains of promises.

There still needs to be a way to organize this process. System designers need a starting point. The obvious starting point is the customer. Continuous design starts by modeling an organization, or any subset of an organization, from the customer in. Starting by mapping the customer's journey helps the organization define the co-creative relationship from the customer's point of view. This definition takes place through a process of exploring questions such as:

- What help do customers need from us in order to accomplish their jobs-to-be-done?

- Where, when, and how do they need to interact with us to get that help?

- What are the relationships between those interaction touchpoints?

- What promises do we need to make at each touchpoint?

This last question brings service design into the realm of promise thinking. Having understood its relationship with the customer, the organization then needs to understand the internal relationships required to keep its customer promises. Service blueprinting identifies the front-stage (customer-facing) and back-stage (behind-the-scenes) activities needed to enable the entire customer journey, along with the relationships between those activities.

Customer journey mapping and service blueprinting address the continuous design principle of designing for service. From that starting point, the entire service organization unfolds through the asking and answering of one key question: How do we maximize our ability to keep our customer promises?

This process repeats itself recursively. Each level in turn defines itself by promising to:

- Design for service, not just software

- Minimize latency and maximize feedback

- Design for failure and operate to learn

- Use operations as input to design

• Seek empathy

Every component at every level of the organization, whether bounded by discipline, microservice pod, or front-stage/back-stage distinction, defines its relationships to other components in terms of the promises they make to one another. Each component defines its boundary as a set of promises through which it supports its customers' journeys. In the case of a microservice API, that boundary could be as simple as a promise to return the phone number associated with a given customer ID. In the case of a design team, it could be as complex as a promise to value outcomes over deliverables.

Each component then defines its internal structure to be whatever is required to keep its customer promises. In many cases, that structure will include dependencies on promises from other components. The phone number microservice might need help from a database service. The design group might need help from the product marketing group. In this way, an organization's service blueprint *inflates* itself in a scalable, resilient manner without reliance on a top-down, centralized, industrial-style process.

Continuous Conversation

The point of continuous design and the promise thinking that drives it is not to invent a newfangled way of doing industrial-era design. Neither customer journey mapping nor service blueprinting are intended to generate big, pretty, plotter-printed artifacts to be hung on walls and used as operational gospel. Taking that approach would go against the entire spirit of continuous quality as continuous repair.

The real value in service design, or any of the other practices that make up the conversational medium, is precisely its power to drive *conversation*. The point of the map or the blueprint is not the thing that results from the process but rather the process that arises from the thing.

Promise thinking treats asking questions as primary and answering them as secondary. It is for this reason that it's called *promise thinking*. The core promise-thinking questions (What promises must we make? What promises do we need? How do we keep our promises?) provide a unifying conversational topic. Their power comes from the way in which they encourage organizations and individuals to think about failure, resilience, and benefit as part of their daily work.

Thinking as espoused by design thinking and promise thinking is not different from doing, but rather encompasses it. It reflects the cybernetic understanding that doing must adapt over time. If design addresses "What should come next?" and operations addresses "What should happen now?", continuous design provides a mechanism for feeding design and operations into each other.

Central to this book is the claim that post-industrialism requires mutual feedback between design and operations. The digital conversational medium works by unifying them within a cybernetic loop. Promise theory provides a single language for designing and executing that loop. Promise thinking is no more nor less than the infusion of this language into a digital service's organizational drinking water. Its emphasis on never-ending questioning is what drives design to become continuous.

Promise thinking is a mindset more than it is a set of procedures. This book does not offer an instruction manual for building a digital conversational medium. It does not contain a checklist for determining whether or not your organization is *promise thinking* correctly or whether it is executing the proper continuous design steps at the right time and in the right order. Its fundamental premise is rather that thinking in promises—continually asking and answering questions about your ability to make and keep promises to customers—*is* the digital conversational medium.

Adopting Promise Thinking

Boiled down to its essence, promise thinking represents the viewpoint that the possibility of failure and the need to continuously validate success must be incorporated within normal operations. Adopting this viewpoint requires transforming your outlook from one of grasping at stability and fearing failure to one of finding resilience by welcoming failure and transforming it into success.

Transforming your outlook is much harder than merely adopting a new tool. Like any new mindset, promise thinking must seep its way through an organization. Especially in the post-industrial era, mandated adoption doesn't scale, and risks cargo-cult implementations that at best are impotent and at worst invite passive resistance.

IT needs a change agent to shepherd the adoption process. QA is ideally positioned to play this role. As the part of IT that most naturally thinks in promises (or, to put it in more QA-centric language, that recognizes the importance of confronting failure to achieve success), QA can lead the rest of the organization to water simply by starting to ask the basic promise-thinking questions as part of their daily work.

Asking whether a service is making the right promises naturally maps to validating requirements. Asking what a service needs to do to keep its promises naturally maps to identifying implementation holes and bugs. Asking what other promises a service needs naturally maps to integration testing. QA's first mission as a boundary-spanning mirror is to help the entire organization begin to see things from a promise-thinking perspective. In order to accomplish this mission, QA need only begin mirroring an organization's promissory nature back to it.

Creating the Digital Conversational Medium

Post-industrialism challenges companies and individuals alike to shift their expectations from consumption to co-creation. It also challenges them to learn how to co-create value in environments that are highly dynamic and uncertain. Digital businesses and the IT organizations that power them need a new conceptual model that helps them engage in the kind of systems thinking required to navigate this new, post-Newtonian economy.

Promise thinking encapsulates the power of cybernetics within a pragmatic and simple, yet comprehensive language. By unifying design and operations into a single model, it captures the continuous design principles of service, feedback, learning, continuity, and empathy. These characteristics make it ideally suited to help post-industrial IT organizations transform themselves into a medium for empathic conversations between digital brands and their customers.

Afterword

The post-industrial economy values service over products, processes over objects, change over stability. The accelerating infusion of physical services with digital components makes it critical for IT to undergo the same transition. The deepening complexity of twenty-first-century life and business, coupled with the increasingly disruptive nature of the market, calls for IT to fully transform itself from a reactive servant of efficiency to a proactive agent of learning.

In order to become a digital conversational medium that enables continuous organizational learning, IT must transcend its perspective on itself as an engineering discipline. IT's new purpose is to help businesses self-steer. To fulfill this purpose, IT must learn to view itself as an agent of design. It must see its role as helping service organizations take Herbert Simon's definition of design to heart by continually changing existing situations into preferred ones.

Delivering Design

Learning happens when internal mechanisms can no longer adequately control external situations. A system learns by reorganizing itself.[1] In a highly dynamic and disruptive environment, learning must become relatively continuous. Continuous learning requires an intimate relationship between design and operations to the point where the boundaries between them begin to blur.

Design addresses the question of what to do next. In order to accurately answer that question, it needs the ability to observe its environment as well as the impact of potential answers. Design thinking uses techniques such as ethnography, prototyping, and user testing for this purpose.

---

1 [ashby1966]

In a complex digital service environment, however, observation through *testing* can only go so far. The most accurate environmental observations come from production operations. According to the cybernetic view, one cannot answer the question of what comes next without feedback from what is happening now. In a competitive market, the business that can accelerate feedback loops by deeply merging design and operations will have the greatest chance of success. Whereas industrial businesses relied on IT to deliver systems, stability, and information, post-industrial businesses rely on post-industrial IT to deliver the capability for continuous design.

John Boyd was a renowned fighter pilot and military strategist for the U.S. Air Force. He developed a highly influential military theory known as the *OODA Loop*. His theory's influence has spread well beyond the confines of the military and is used by business and technology strategists across many industries.

The acronym OODA stands for Observe-Orient-Decide-Act. In order to shoot down an enemy aircraft or even just respond to its maneuvers without being shot down oneself, a pilot must observe the enemy, identify a set of possible actions based on this observation, pick one, and execute it. Boyd believed that a fighter pilot who could mentally navigate the OODA loop more quickly than the enemy would gain the upper hand in battle.

Design and operations constitute overlapping OODA loops as part of the digital conversational medium. On the one hand, designers observe, orient, decide, and act during the process of designing a new solution. On the other hand, operations staff observe, orient, decide, and act during the process of monitoring and responding to problems with production environments.

By generating changes that must be deployed, designers' actions trigger operational actions. Conversely, operational observations contribute to design observations by providing insights into real system and customer behavior. An optimal conversational medium deeply integrates them with each other and minimizes the friction that distinguishes them. The ultimate expression of the digital conversational medium is an environment where everyone understands and contributes to design and operations.

Designing Delivery

This book has introduced concepts such as cybernetics, autopoiesis, and self-steering that might be unfamiliar to many readers. The basic premise behind these concepts is that *being* is defined by *becoming*. My purpose in using them is to express the view that post-industrial business requires a digital medium that enables continual change, adaptation, and learning. This medium must make the continual design of services, as well as the organizations that operate those services, natural and fluid. Second-order cybernetics provides a powerful lens through which to understand these principles and their implications for how IT organizations see themselves.

If change is the only constant, then design is inseparable from operations, and thus never-ending. This dictum holds for IT's view of itself as much as for its view of its role within its surrounding business. Specific IT practices are less important than the continual evolution of how we use them. Whether one follows LeanUX as laid out in Jeff Gothelf's book,[2] for example, or whether one even calls what one does LeanUX, is beside the point.

What matters is how well IT keeps its promise to help digital businesses maintain brand quality through digital conversations with their customers. In order for the business to self-steer as part of that conversation, IT also needs to self-steer. It needs to treat any particular methodology as a provisional, momentary solution for keeping its promises.

IT needs to treat new methodologies as opportunities for cybernetic conversations: What does this practice mean to us? How do we adopt it? How do we adapt it and ourselves to each other? In the post-industrial economy, IT's new purpose is to enable digital businesses to continually redesign themselves. To fulfill that purpose, IT also needs to continually redesign itself as a delivery mechanism. It needs to transform its understanding of itself from a *thing* to a *process*. Furthermore, it needs to begin to act from a deep understanding that the continually unfolding process that is IT is inseparable from the continually unfolding process that is the surrounding business.

Digital business as a circular process of empowering and responding to customers is the fundamental raison d'être for post-industrial IT. It is the essence of promise thinking. In the quest for continuous quality, we use strange words like cybernetics, autopoiesis, and self steering. We define strange languages like promise theory and second-order analysis. We do so because we face a strange new world of service, infusion, complexity, and disruption.

Customers and businesses need IT's help to steer their way through these new surroundings that are so much choppier and more winding than the ones to which they're accustomed. The point of the digital conversational medium and everything we do to design and operate it is nothing more nor less than maximiz-

---

2 [gothelf2013]

ing our ability to actualize our own empathy. By doing so, we can help the customers and businesses we serve steer their way toward success and satisfaction.

Bibliography

Cybernetics

- [ashby1966] W. Ross Ashby, *Design for a Brain* (London: Chapman and Hall, 1966).

- [ashby1957] W. Ross Ashby, *An Introduction to Cybernetics* (London: Chapman and Hall Ltd., 1957).

- [bateson2002] Gregory Bateson, *Mind and Nature: A Necessary Unity* (New York: Hampton Press, 2002).

- [conwaysiegelman2006] Flo Conway and Jim Siegelman. *Dark Hero of the Information Age* (New York: Basic Books, 2006).

- [maturanavarela1980] Humberto R. Maturana and Francisco Varela. *Autopoiesis and Cognition: The Realization of the Living* (Dordrecht, Netherlands: D. Reidel Publishing Co., 1980).

- [maturana1992] Humberto R. Maturana, *The Tree of Knowledge* (Boston: Shambhala, 1992).

- [medina2014] Eden Medina, *Cybernetic Revolutionaries* (Cambridge, MA: The MIT Press, 2014).

- [rosenbluth1943] Arturo Rosenbluth, Norbert Wiener, and Julian Bigelow, "Behavior, Purpose, and Teleology," *Philosophy of Science* 10 (1943: 18–24).

- [vonfoerster2003] Heinz von Foerster, *Understanding Understanding: Essays on Cybernetics and Cognition* (New York: Springer-Verlag, 2003).

- [wiener1965] Norbert Wiener, *Cybernetics: or Control and Communication in the Animal and the Machine* (Cambridge, MA: MIT Press, 1965).

- [wiener1954] Norbert Wiener. *The Human Use of Human Beings* (New York: Doubleday and Co., 1954).

- [wiener1954_2] Norbert Wiener. "The Role of the Observer," *Philosophy of Science* 3 (1936): 307–319.

- [winogradflores1987] Terry Winograd and Fernando Flores. *Understanding Computers and Cognition* (Upper Saddle River, NJ: Addison-Wesley, 1987).

Complexity and Systems Thinking

- [burgess2014] Jan A. Bergstra and Mark Burgess. *Promise Theory: Principles and Applications* (Oslo, Norway: χtAxis Press, 2014).

- [burgess2015] Mark Burgess. *In Search of Certainty* (Sebastopol, CA: O'Reilly, 2015).

- [dekker2011] Sydney Dekker. *Drift into Failure* (Burlington, VT: Ashgate Publishing Co., 2011).

- [gellman1995] Murray Gell-Man. *The Quark and the Jaguar* (New York: St. Martin's Griffin, 1995).

- [macy1991] Joanna Macy, *Mutual Causality in Buddhism and General Systems Theory* (New York: State University of New York Press, 1991).

- [meadows2008] Donella H. Meadows, *Thinking in Systems: A Primer* (White River Junction, VT: Chelsea Green Publishing, 2008).

- [osinga2007] Frans P.B. Osinga, *Science, Strategy, and War* (New York: Routledge, 2007).

- [varelathompsonrosch1992] Francisco Varela, Evan Thompson, and Eleanor Rosch. *The Embodied Mind* (Cambridge, MA: MIT Press, 1992).

- [vonbertalanffy1969] Ludwig von Bertalanffy, *General System Theory* (New York: George Braziller Inc., 1969).

Design Thinking

- [brown2009] Tim Brown. *Change by Design: How Design Thinking Transforms Organizations and Inspires Innovation* (New York: HarperBusiness, 2009).

- [krippendorff2007] Klaus Krippendorff, "The Cybernetics of Design and the Design of Cybernetics," *Kybernetes* 36 (9/10; 2007): 1381–1392.

- [lim2013] Seung Chan Lim. *Realizing Empathy: An Inquiry into the Meaning of Making* (2013).

- [simon1996] Herbert A. Simon, *The Sciences of the Artificial, Third Edition* (Cambridge, MA: MIT Press, 1996).

- [stickdorn2010] Mark Stickdorn and Jakob Schneider. *This Is Service Design Thinking* (Amsterdam: BIS Publishers, 2010).

- [wendt2015] Thomas Wendt. *Design for Dasein* (2015).

Lean and DevOps

- [busche2014] Laura Busche. *Lean Branding* (Sebastopol, CA: O'Reilly, 2014).

- [hendrickson2013] Elizabeth Hendrickson, *Explore It!* (Frisco, TX: Pragmatic Bookshelf, 2013).

- [humblefarley2010] Jez Humble and David Farley. *Continuous Delivery* (Upper Saddle River, NJ: Addison-Wesley, 2010).

- [gothelf2013] Jeff Gothelf, *Lean UX* (Sebastopol, CA: O'Reilly, 2013).

- [kimbehrspafford2014] Gene Kim, Kevin Behr, and George Spafford. *The Phoenix Project* (IT Revolution Press, 2014).

- [whittakerarboncarollo2012] James Whittaker, Jason Arbon, and Jeff Carollo, *How Google Tests Software* (Upper Saddle River, NJ: Addison-Wesley, 2012).

Management and Economics

- [bell1976] Daniel Bell, *The Coming of Post-Industrial Society* (New York: Basic Books, 1976).

- [christensenraynor2003] Clayton M. Christensen and Michael E. Raynor, *The Innovator's Solution* (Cambridge, MA: Harvard Business School Press, 2003).

- [gray2012] Dave Gray, *The Connected Company* (Sebastopol, CA: O'Reilly, 2012).

- [taylor1997] Frederick Winslow Taylor, *The Principles of Scientific Management* (New York: Dover Publications, 1997).

- [vargolusch2004] Steven L. Vargo and Robert F. Lusch, "Evolving to a New Dominant Logic for Marketing." *Journal of Marketing, 68* (1–17, 2004).

- [vargolusch2014] Steven L. Vargo and Robert F. Lusch, *Service-Dominant Logic* (Cambridge, UK: Cambridge University Press, 2014).

Index

Q

R

About the Author

Jeff Sussna is Founder and Principal of Ingineering.IT, a Minneapolis consulting firm that helps companies adopt post-industrial IT practices. Jeff has nearly 30 years of IT experience. He has led high-performance teams across the development/QA/operations spectrum. He specializes in driving quality improvements through practical innovation. Jeff has done work for a diverse range of companies, including Fortune 500 enterprises, major technology companies, software product and service startups, and media conglomerates.

Jeff combines engineering expertise with the ability to bridge business, creative, and technical perspectives. He has the insight and experience to uncover problems and solutions others miss. He is a highly sought-after speaker and writer respected for his insights on topics such as Agile, DevOps, service design, and cloud computing. His interests focus on the intersection of development, operations, design, and business.

Colophon

The cover font is Gotham. The text font is Scala Pro Regular; the heading font is Benton Sans; and the code font is Dalton Maag's Ubuntu Mono.

Get even more for your money.

Join the O'Reilly Community, and register the O'Reilly books you own. It's free, and you'll get:

- $4.99 ebook upgrade offer
- 40% upgrade offer on O'Reilly print books
- Membership discounts on books and events
- Free lifetime updates to ebooks and videos
- Multiple ebook formats, DRM FREE
- Participation in the O'Reilly community
- Newsletters
- Account management
- 100% Satisfaction Guarantee

Signing up is easy:

1. Go to: oreilly.com/go/register
2. Create an O'Reilly login.
3. Provide your address.
4. Register your books.

Note: English-language books only

To order books online:
oreilly.com/store

For questions about products or an order:
orders@oreilly.com

To sign up to get topic-specific email announcements and/or news about upcoming books, conferences, special offers, and new technologies:
elists@oreilly.com

For technical questions about book content:
booktech@oreilly.com

To submit new book proposals to our editors:
proposals@oreilly.com

O'Reilly books are available in multiple DRM-free ebook formats. For more information:
oreilly.com/ebooks

O'REILLY®

9 781491 949887